Lecture Notes in Mathematics

Edited by A. Dold and B. Eckmann

Subseries: Mathematisches Institut der Universität und
Max-Planck-Institut für Mathematik, Bonn – vol. 9
Adviser: F. Hirzebruch

1244

Werner Müller

Manifolds with Cusps of Rank One

Spectral Theory and L^2-Index Theorem

Springer-Verlag
Berlin Heidelberg New York London Paris Tokyo

Author

Werner Müller
Akademie der Wissenschaften der DDR
Karl-Weierstraß-Institut für Mathematik
Mohrenstraße 39, DDR – 1086 Berlin, German Democratic Republik

Mathematics Subject Classification (1980): 58G10, 58G11, 58G25

ISBN 3-540-17696-9 Springer-Verlag Berlin Heidelberg New York
ISBN 0-387-17696-9 Springer-Verlag New York Berlin Heidelberg

Printing and binding: Druckhaus Beltz, Hemsbach/Bergstr.
2146/3140-543210

INTRODUCTION

Let G be a connected real semisimple Lie group of noncompact type, K a maximal compact subgroup of G and G/K the associated globally symmetric space. Consider a discrete torsion-free subgroup Γ of G with finite covolume and let $\Gamma\backslash G/K$ be the corresponding locally symmetric space. Let V,W be finite-dimensional unitary K-modules and denote by $\tilde{E}$, $\tilde{F}$ the induced homogeneous vector bundles over G/K. $\tilde{E}$ and $\tilde{F}$ can be pushed down to locally homogeneous vector bundles $E = \Gamma\backslash\tilde{E}$ and $F = \Gamma\backslash\tilde{F}$ over $\Gamma\backslash G/K$. Let

$$\tilde{\mathcal{D}}: C^\infty(G/K,\tilde{E}) \longrightarrow C^\infty(G/K,\tilde{F})$$

be an invariant elliptic differential operator. Then $\tilde{\mathcal{D}}$ induces an elliptic differential operator

$$\mathcal{D}: C^\infty(\Gamma\backslash G/K,E) \longrightarrow C^\infty(\Gamma\backslash G/K,F) .$$

It is proved in [61] that $\mathcal{D}$ has a well-defined L^2-index which, as in the compact case, depends only on $\mathrm{ch}\,V - \mathrm{ch}\,W$. Using Selberg's trace formula, Barbasch and Moscovici [15] derived an explicit formula for the L^2-index of $\mathcal{D}$ if the locally symmetric space $\Gamma\backslash G/K$ has strictly negative curvature or, equivalently, if the real rank of G equals one. It seems to be very interesting to have an explicit formula for the L^2-index in the general case.

In this book we shall investigate the case of a locally symmetric space of $\mathbb{Q}$-rank one. Actually, we shall work with a larger class of manifolds. Each of these manifolds is locally symmetric near infinity with ends generalizing the case of a cusp of a $\mathbb{Q}$-rank one locally symmetric space. This is motivated by our approach to the proof of a conjecture of Hirzebruch (c.f. [63,§6]). In [63] we investigated the signature operator on Hilbert modular varieties $X = \Gamma\backslash H^n$. Here $\Gamma = SL(2,\mathcal{O}_F)$ is the Hilbert modular group of a totally real number field F of degree n. The L^2-index of the signature operator was computed with the help of the Selberg trace formula. The contribution of the cusps in the index formula was given by special values of certain L-series associated to the number field F. These special values of the

L-series occure in the formula conjectured by Hirzebruch relating signature defects of cusps of Hilbert modular varieties and special values of L-series. We proved that for a Hilbert modular variety with a single cusp Hirzebruch's conjecture is a consequence of our index formula. This suggests that an explicit formula for the L^2-index of the signature operator on manifolds investigated in this book will have other interesting applications of this type.

A proof of Hirzebruch's conjecture was given by Atiyah, Donnelly and Singer in [6] . This proof is based on the former work of Atiyah Patodi and Singer [7] on spectral asymmetry.

In the present book we shall give another proof of Hirzebruch's conjecture along the lines briefly sketched in [63,§6] . Actually this will turn out to be one application of the index formula we shall establish in this book.

As indicated above, in this book we are dealing with manifolds which are locally symmetric near infinity with ends generalizing the case of $\mathbb{Q}$-rank one cusps. On manifolds of this type we shall investigate a class of first-order elliptic differential operators which we call **generalized chiral Dirac operators**. One of our purposes is to establish a formula for the L^2-index of these operators. This covers the case of twisted Dirac operators on $\mathbb{Q}$-rank one locally symmetric spaces. It is known that this is sufficient to compute the L^2-index of any locally invariant elliptic differential operator on a $\mathbb{Q}$-rank one locally symmetric space (c.f. [15,p.196]). The main contribution of the cusps in this index formula is again given by a special value of a certain L-series associated to the locally symmetric structure of the ends of the manifolds. This generalizes the L-series arising in the Hilbert modular case.

We shall now give a more detailed description of the content of this book. In §1 we have collected some auxiliary results from the theory of linear operators in Hilbert space. We recall here some results of the Krein-Birman theory of the spectral shift function and also some facts concerning supersymmetric scattering theory [82] . In §2 we introduce the cusps we shall consider in this book. Each cusp is a locally symmetric space $Y = \Gamma\backslash G/K$ of infinite volume. Y is diffeomorphic to a cylinder $\mathbb{R}^+ \times \Gamma\backslash Z$ where Z is a certain homogeneous space and $\Gamma\backslash Z$ is compact. Moreover, $\Gamma\backslash Z$ is a fibration over a compact locally symmetric space $\Gamma_M\backslash X_M$ with fibre a compact nilmanifold. For each $b \geq 0$, we denote by Y_b the submanifold of Y which corresponds to $[b,\infty) \times \Gamma\backslash Z$. Each submanifold Y_b, $b \geq 0$, of Y with the induced Riemannian metric will be called a **cusp of rank one**. In §3 we study the fundamental solution of the heat equation for certain locally invariant differential operators on the cusp Y. For the same kind of locally invariant differential operators on Y we investigate in §4 the Neumann problem on the submanifolds Y_b, $b > 0$. These results are basic for §§5 and 6. In §5 we consider manifolds with cusps of rank one. Such a manifold is a complete Riemannian manifold X which is the union of a compact manifold with boundary and a finite number of cusps of rank one. For simplicity we shall assume throughout this book that X has a single cusp. Thus $X = X_o \cup Y_1$ where X_o is a compact manifold with boundary, Y_1 is a cusp of rank one and $X_o \cap Y_1 = \partial X_o = \partial Y_1$. The extension of our results to manifolds with several cusps requires nothing which is essentially new. On X we shall consider differential operators D which are locally invariant at infinity, i.e., there exists a locally invariant differential operator on the cusp Y whose restriction to Y_1 coincides with the restriction of D to Y_1. To be able to apply harmonic analysis at infinity we consider a restricted class of differential operators which we call **generalized Dirac operators**. Let E be a complex vector bundle over X whose restriction to Y_1 coincides with the restriction to Y_1 of a certain locally homogeneous vector bundle over Y. A **generalized Dirac operator** is a first-order formally selfadjoint elliptic differential operator D on $C^\infty(X,E)$ which is locally invariant at infinity and such that D^2 coincides, up to a zero-order operator, with the operator induced by the Casimir operator of G on $C^\infty(Y_1,E|Y_1)$. Geometrically interesting operators are of this form. In §5 we shall prove that D^2 acting in $L^2(X,E)$ with domain $C_c^\infty(X,E)$ is essentially selfadjoint. Let H be the unique selfadjoint extension of D^2. In §6 we shall investigate the spectral resolution of H. For this purpose we introduce an auxiliary operator H_o whose continuous spectrum can be explicitly described such that $(H + I)^{-1}$ is a compact perturbation of $(H_o + I)^{-1}$. H_o is obtained from H by imposing Neumann boundary conditions on the hypersurface ∂Y_2, $Y_2 \subset Y$ as above. Employing the results of §4 we determine the continuous spectrum of H_o. Then we prove that the wave operators $W_\pm(H,H_o)$ exist and are complete which implies that the absolutely continuous parts of H and H_o are unitarily equivalent. To establish the existence and completeness of the wave operators we employ the method introduced by Enss [31] in quantum scattering theory. Actually we shall apply an abstract version of this method introduced by Amrein, Pearson and Wollenberg [2], [16]. This method gives even more - the absence of the singularly continuous spectrum of H. This is Theorem 6.17. Then we continue with the study of the eigenvalues of H. We employ the method of Donnelly [28], which he used in the case of $\mathbb{Q}$-rank one locally symmetric spaces. There is no problem to extend this method to our case. The result is that the number of eigenvalues of H which are less than λ, $\lambda > 0$, is bounded by a constant multiple of λ^d for a certain $d \in \mathbb{N}$. Let H_d be the restriction of H to the subspace spanned by the eigenfunctions of H. Then our estimate on the growth of the number of eigenvalues implies that, for each $t > 0$, $\exp(-tH_d)$ is of the trace class. Another consequence is that $\ker D \cap L^2$ is finite-dimensional. Therefore each generalized chiral Dirac operator $D: C^\infty(X,E^+) \longrightarrow C^\infty(X,E^-)$ has a finite L^2-index, denoted $L^2\text{-Ind}\, D$. Let H^+ (resp. H^-) be the unique selfadjoint extensions of D^*D (resp. DD^*). Then the results of §6 imply

$$L^2\text{-Ind}\, D = \operatorname{Tr}(\exp(-tH_d^+)) - \operatorname{Tr}(\exp(-tH_d^-)) \quad . \tag{0.1}$$

In §7 we construct the kernel $e(z,z',t)$ of the heat operator $\exp(-tH)$. For this purpose we employ a variant of the usual parametrix method as in [62]. To construct the parametrix at infinity we apply the results of §3. In §8 we construct a system of generalized eigenfunctions for the operator H. In the locally symmetric case such a system of generalized eigenfunctions is given by the **Eisenstein series**. We call the generalized eigenfunctions in the general case **Eisenstein functions.** The proof of the existence of an analytic continuation of the Eisenstein functions is due to L.Guillopé [38]. Using the Eisenstein functions we get an explicit description of the wave operators $W_\pm(H,H_o)$. Together with the results of §6 we recover in this way all facts known about the spectral resolution of the Casimir operator acting on sections of a locally homogeneous vector bundle over a $\mathbb{Q}$-rank one locally symmetric space. In §9 we investigate the spectral shift function $\xi(\lambda;H,H_o)$ associated to H and H_o. The main result is Theorem 9.25 which gives an explicit expression for the spectral shift function $\xi^c(\lambda;H,H_o)$ associated to the absolutely continuous parts H_{ac} and $H_{o,ac}$ of H and H_o, respectively. These results are used in §10 to derive a preliminary version of our index formula for a generalized chiral Dirac operator

$$D: C^\infty(X,E^+) \longrightarrow C^\infty(X,E^-) \quad .$$

We work within the supersymmetric framework developed in [82]. Thus we regard $\begin{pmatrix} 0 & D^* \\ D & 0 \end{pmatrix}$ as an operator in $L^2(X,E^+) \oplus L^2(X,E^-)$ with domain given by $C_c^\infty(X,E^+) \oplus C_c^\infty(X,E^-)$ and we denote by Q its unique selfadjoint extension. Set $H = Q^2$. Then we introduce a free Hamiltonian $\tilde{H}_o$ which is a modification of the free Hamiltonian H_o considered in §6. To define it recall that the boundary of the cusp Y_1 is a fibre bundle over a compact locally symmetric space $\Gamma_M\backslash X_M$. If we restrict D to sections of $E^+|Y_1$ which are constant along the fibres then we get an operator

$$\mathcal{D}_o\colon C^\infty(\mathbb{R}^+ \times \Gamma_M\backslash X_M, E_M^+) \longrightarrow C^\infty(\mathbb{R}^+ \times \Gamma_M\backslash X_M, E_M^-)$$

which is of the form

$$\mathcal{D}_o = \eta\left(\frac{\partial}{\partial u} + \tilde{D}_M\right) \quad .$$

Here η is a bundle isomorphism and $\tilde{D}_M$ is a selfadjoint operator in $C^\infty(\Gamma_M\backslash X_M, E_M^+)$. Let Q_o^+ be the closure in L^2 of $\partial/\partial u + \tilde{D}_M$ with respect to the non local boundary conditions of [7] and let Q_o^- be its Hilbert space adjoint. This is the closure in L^2 of $-\partial/\partial u + \tilde{D}_M$ with respect to the adjoint boundary conditions. Then $Q_o = \begin{pmatrix} 0 & Q_o^- \\ Q_o^+ & 0 \end{pmatrix}$ is a selfadjoint operator in $L^2(\mathbb{R}^+ \times \Gamma_M\backslash X_M, E_M^+ \oplus E_M^+)$ and our free Hamiltonian is given by $\tilde{H}_o = Q_o^2$. Then (Q,Q_o) define a supersymmetric scattering theory in $L^2(X,E^+ \oplus E^-)$ (c.f. Ch.I). Therefore, we may split the spectral shift function $\xi^c(\lambda;H,H_o)$ into a bosonic part $\xi_+^c(\lambda;H,H_o)$ and a fermionic part $\xi_-^c(\lambda;H,H_o)$. Further, let $e_+(z,z',t)$ (resp. $e_-(z,z',t)$) be the bosonic (resp. fermionic) part of the heat kernel $e(z,z',t)$ for the Hamiltonian H. Then our first result is

$$\begin{aligned} L^2\text{-Ind}\, D = {} & \int_X \{\operatorname{tr} e_+(z,z,t) - \operatorname{tr} e_-(z,z,t)\}dz \; + \\ & + \sum_\omega \frac{\operatorname{sign}}{2} \operatorname{erfc}(|\omega|\sqrt{t}) \; + \qquad\qquad (0.2) \\ & + t\int_0^\infty (\xi_+^c(\lambda;H,H_o) - \xi_-^c(\lambda;H,H_o))e^{-t\lambda}d\lambda \quad . \end{aligned}$$

Here ω runs over the eigenvalues of $\tilde{D}_M$ and erfc is the complementary error function. Using Theorem 9.25 and results of [82] , it follows that $\xi_+^c(\lambda;H,H_o) - \xi_-^c(\lambda;H,H_o)$ is constant on $\mathbb{R}^+$. This constant is zero if the lower bound of the essential spectrum of H is positive. According to [7], the second term on the right hand side of (0.2) has an asymptotic expansion (as $t \longrightarrow 0$) whose constant term is $\frac{1}{2}(\eta(0)+h)$ where $\eta(0)$ is the Eta invariant of $\tilde{D}_M$ and

$h = \dim \ker \tilde{D}_M$. Let $\alpha^{\pm}(z)$ be the constant term in the asymptotic expansion of $\operatorname{tr} e_{\pm}(z,z,t)$ as $t \to 0$. Then the local index theorem [5], [34], leads to our first version of an index formula

$$L^2\text{-Ind}\, D = \int_X (\alpha^+(z) - \alpha^-(z))dz + U + \frac{1}{2}\eta(0) - \frac{1}{2}(h_\infty^+ - h_\infty^-) . \tag{0.3}$$

The term U is determined by the asymptotic expansion of the heat kernel at infinity. We call U the **unipotent contribution** to the L^2-index of D. $h_\infty^{\pm}$ are the dimensions of spaces of extended L^2-solutions of D and D*, respectively, with limiting values in $\ker \tilde{D}_M$. In particular, the last term vanishes if the lower bound of the essential spectrum of H is positive or, equivalently, if the continuous extension $\bar{D}: H^1(X,E^+) \longrightarrow L^2(X,E^-)$ of D is a Fredholm operator. Of course the index formula (0.3) is not of much use unless the unipotent contribution U has been made more explicit. We deal with this problem in §11. In the case of Hilbert modular varieties we proved in [63] that the unipotent contribution to the L^2-index of the signature operator is given by the value at s=1 of a certain L-series associated to the cusp (c.f. [63,(5.57)]). We shall show that in the general case one has in principle the same picture. In the first part of §11 we reduce the computation of the unipotent contribution to the study of the asymptotic expansion (as $t \longrightarrow 0$) of the integral (11.39). Using ideas of a forthcoming paper of W.Hoffmann [51] the integral (11.39) is then converted into a finite sum of unipotent orbital integrals and a certain non-invariant integral (c.f. (11.56)). In this book we shall not investigate the unipotent orbital integrals and the non-invariant integral occurring in (11.56) in general. This requires the knowledge of a Fourier inversion formula for the corresponding distributions on G. A Fourier inversion formula for unipotent orbital integrals for groups of real rank one was established by Barbasch [14]. If $G = G_1 \times \cdots \times G_r$ where each G_i is a connected semisimple Lie group of real rank one then one can employ the results of Barbasch to compute the unipotent orbital integrals in our case. Under the same assumption one can also deal with the non-invariant integral in (11.56). This leads to our final index formula for the case where $G = G_1 \times \cdots \times G_r$ with G_i as above (c.f. Theorem 11.77). If $\operatorname{rank} G > \operatorname{rank} K$, then the unipotent contribution U vanishes. If $\operatorname{rank} G = \operatorname{rank} K$, then there is an L-series associated to the cusp and the differential operator D such that the unipotent contribution U to the L^2-index of D is given by the value at $s=0$ of this L-series. This generalizes our results concerning the L^2-index of the signature operator on Hilbert modular varieties obtained in [63]. There is no doubt that a similar result will be true without any restriction on G. Moreover, if the group Γ arises from an arithmetic situation then the L-series introduced by (11.67b) should have an arithmetic meaning.

In this book we shall discuss only one application of our index formula. This is the proof of the Hirzebruch conjecture which will be derived from our index formula in §12. It is clear that this method can be generalized to other locally symmetric spaces. This will lead to a generalization of Hirzebruch's conjecture which can be proved along the same lines.

I thank L.Guillopé for pointing out mistakes in the first draft of this book.

It is a pleasure for me to thank Mme. Breiner (I.H.E.S.) and Mme. Wüst (Berlin) who typed parts of the manuscript.

TABLE OF CONTENTS

CHAPTER I
PRELIMINARIES

For the convenience of the reader we shall collect here some auxiliary results from the theory of linear operators in Hilbert space.

Let $\mathcal{H}$ be a separable Hilbert space. The domain of any operator T in $\mathcal{H}$ will be denoted by $\mathbb{D}(T)$. Let T be a symmetric positive operator in $\mathcal{H}$ with dense domain $\mathbb{D}(T)$. On $\mathbb{D}(T)$ we define a new scalar product by

$$(f,g)_T = (f,g) + (Tf,g) \quad , \quad f,g\in\mathbb{D}(T) \quad .$$

Following Friedrichs [33] , introduce the subspace $\mathbb{D}[T] \subset \mathcal{H}$ by

$$\mathbb{D}[T] = \{ \ f\in\mathcal{H} \mid \exists \{f_n\}_{n\in\mathbb{N}} \subset \mathbb{D}(T) \ \text{ such that } \ \|f_n - f\| \to 0 \ \text{ as } \ n\to\infty \ \text{ and } \ \|f_n - f_m\|_T \longrightarrow 0 \text{ as } n,m \longrightarrow \infty \ \} .$$

The norm $\|\cdot\|_T$ can be extended to a norm on $\mathbb{D}[T]$. $\mathbb{D}[T]$ equipped with $\|\cdot\|_T$ is a Hilbert space and $\mathbb{D}(T)$ is a dense subspace. By $\tilde{T}_o$ we shall denote Friedrichs' extension of T [33]. $\tilde{T}_o$ is a positive selfadjoint operator in $\mathcal{H}$. Its domain is given by

$$\mathbb{D}(T_o) = \mathbb{D}[T] \cap \mathbb{D}(T^*)$$

and $\tilde{T}_o = T^* | \mathbb{D}(\tilde{T}_o)$. $\tilde{T}_o$ is the unique positive selfadjoint extension of T which satisfies $\mathbb{D}(\tilde{T}_o) \subset \mathbb{D}[T]$. Moreover, one has $\mathbb{D}[\tilde{T}_o]=\mathbb{D}[T]$, [1, No 109,Theorem 2], [30,XII,§5] . Another way to define Friedrichs' extension is the following one. Consider the quadratic form

$$q(f) = (Tf,f) \quad , \quad f\in\mathbb{D}(T) \quad ,$$

and let $\bar{q}$ be its closure in $\mathcal{H}$. There exists a salfadjoint operator T' in $\mathcal{H}$ which represents the quadratic form $\bar{q}$. This operator coincides with Friedrichs' extension $\tilde{T}_o$ defined above [52,VI,§2] .

Now we recall briefly some results of the Krein-Birman theory of the spectral shift function [18], [86] . Let H,H_o be bounded selfadjoint operators in the Hilbert space $\mathcal{H}$ and assume that $H - H_o$ is of the trace class. Then the function

$$\xi(\lambda) = \xi(\lambda;H,H_o) = \pi^{-1}\lim_{\varepsilon\downarrow 0} \arg\det[1 + (H - H_o)(H_o - \lambda - i\varepsilon)^{-1}]$$

exists for a.e. $\lambda\in\mathbb{R}$. This is the spectral shift function associated to H and H_o. It has the following properties:

i) $\xi\in L^1(\mathbb{R})$ with $\|\xi\|_{L^1} \leq \|H - H_o\|_1$ where $\|\cdot\|_1$ is the trace norm.

ii) $\mathrm{Tr}(H - H_o) = \int_{-\infty}^{\infty} \xi(\lambda)d\lambda$.

iii) $\xi(\lambda) = 0$ outside the smallest interval containing the spectra of of H and H_o.

iv) Let $\Phi\in C_o^\infty(\mathbb{R})$. Then $\Phi(H) - \Phi(H_o)$ is of the trace class and

$$\mathrm{Tr}(\Phi(H) - \Phi(H_o)) = \int_{-\infty}^{\infty} \Phi'(\lambda)\xi(\lambda)d\lambda \; .$$

Let $\mathcal{H}^{ac}$ and $\mathcal{H}_o^{ac}$ be the absolutely continuous subspaces of H and H_o , respectively, and denote by P^{ac} and P_o^{ac} the corresponding orthogonal projections of $\mathcal{H}$ onto $\mathcal{H}^{ac}$ and $\mathcal{H}_o^{ac}$, respectively. Since $H - H_o$ is of the trace class, it follows that the wave operators

$$W_\pm(H,H_o) = \operatorname*{s-lim}_{t\to\pm\infty} e^{itH}e^{-itH_o}P_o^{ac}$$

exist and are complete, i.e., $W_\pm(H,H_o)$ is an isometry of $\mathcal{H}_o^{ac}$ onto $\mathcal{H}^{ac}$ which intertwines the absolutely continuous parts $H_{o,ac} = H_o|\mathcal{H}_o^{ac}$ and $H_{ac} = H|\mathcal{H}^{ac}$ of H_o and H, respectively (c.f. [52,X,§4]). Then the scattering operator S is defined by

$$S = W_+^*W_- \; .$$

S is a unitary operator on $\mathcal{H}_o^{ac}$ which commutes with $H_{o,ac}$. Let $dE_{ac}(\lambda)$ be the spectral measure of $H_{o,ac}$. Then there is a corresponding spectral decomposition for S:

$$S = \int S(\lambda)dE_{ac}(\lambda)$$

where $S(\lambda)$ is a bounded operator in the Hilbert space $\mathcal{H}(\lambda) = \frac{dE_{ac}(\lambda)}{d\lambda}\mathcal{H}$. $S(\lambda)$ is the scattering matrix of H,H_o. It is related to the spectral shift function $\xi(\lambda)$ by

$$\exp(-2\pi i\xi(\lambda)) = \det S(\lambda)$$

for a.e. $\lambda \in \sigma_{ac}(H_o)$ (c.f.[16,V,19.1.5]).

Finally, we shall discuss some results from [82] on supersymmetric scattering theory. Assume that τ is a unitary involution in the Hilbert space $\mathcal{H}$. The ± 1 eigenspaces $\mathcal{H}_\pm$ of $\mathcal{H}$ are called the bosonic and fermionic subspaces, respectively. A selfadjoint operator Q in $\mathcal{H}$ is called a **supercharge** with respect to τ if

$$\tau Q = - Q\tau \quad \text{on} \quad \mathbb{D}(Q) .$$

The selfadjoint operator $H = Q^2 \geqq 0$ is called the associated Hamiltonian. Any operator H of this form for some Q and τ is called a Hamiltonian with supersymmetry.

A supersymmetric scattering theory in a Hilbert space $\mathcal{H}$ with a unitary involution τ is given by a pair (Q,Q_o) with the following properties:

i) Q and Q_o are supercharges with respect to τ.

ii) Let $H = Q^2$ and $H_o = Q_o{}^2$ be the associated Hamiltonians. Then the wave operators

$$W_\pm(H,H_o) = \operatorname*{s-lim}_{t \to \pm\infty} e^{itH} e^{-itH_o} P_o^{ac}$$

exist and are complete. Here P_o^{ac} denotes the orthogonal projection of $\mathcal{H}$ onto the absolutely continuous subspace $\mathcal{H}_{o,ac}$ of H_o.

iii) $W_\pm(H,H_o)$ are intertwining operators for Q and Q_o:

$$QW_\pm(H,H_o) = W_\pm(H,H_o)Q_o \quad \text{on} \quad \mathbb{D}(Q_o) \cap \mathcal{H}_{o,ac} .$$

A sufficient condition for the existence of a supersymmetric scattering theory is given by

<u>LEMMA 1.1</u>. Assume that Q and Q_o are supercharges in $\mathcal{H}$ with respect to τ and

$$Q\exp(-tQ^2) - Q_o\exp(-tQ_o{}^2)$$

is of the trace class for all $t > 0$. Then Q and Q_o define a supersymmetric scattering theory in $\mathcal{H}$.

For the proof see [82].

Assume that (Q,Q_o) define a supersymmetric scattering theory with respect to τ . Let

$$Q_o|\mathcal{H}_o^{ac} = \int Q_o(\lambda)dE_{ac}(\lambda)$$

be the spectral decomposition of $Q_o|\mathcal{H}_o^{ac}$ with respect to the spectral measure $dE_{ac}(\lambda)$ of $H_{o,ac}$. Since H_o commutes with τ it follows that

$$\tau = \int \tau(\lambda)dE_{ac}(\lambda)$$

and the Hilbert space $\mathcal{H}(\lambda) = \frac{dE_{ac}(\lambda)}{d\lambda}\mathcal{H}$ admits a decomposition

$$\mathcal{H}(\lambda) = \mathcal{H}_+(\lambda) \oplus \mathcal{H}_-(\lambda) \tag{1.2}$$

into the ± 1 eigenspaces of $\tau(\lambda)$. With respect to this decomposition we may write

$$Q_o(\lambda) = \begin{pmatrix} 0 & q_-(\lambda) \\ q_+(\lambda) & 0 \end{pmatrix} .$$

Now observe that S commutes with τ too, and, by iii), S commutes also with Q_o. This implies that with respect to the decomposition (1.2) we have

$$S(\lambda) = \begin{pmatrix} S_+(\lambda) & 0 \\ 0 & S_-(\lambda) \end{pmatrix}$$

and

$$q_+(\lambda)S_+(\lambda) = S_-(\lambda)q_+(\lambda).$$

CHAPTER II
CUSPS OF RANK ONE

The manifolds we shall consider in this book are locally symmetric near infinity with cusp-like ends. In this chapter we shall describe the locally symmetric spaces which occur as ends of our manifolds and we establish the assumption and notation.

Let G be a connected noncompact real semisimple Lie group with finite center. G is admissible in the sense of [66, Ch.2]. Let K be a maximal compact subgroup of G and set $\tilde{Y} = G/K$. The Lie algebras of G and K are denoted by $\mathfrak{g}$ and $\mathfrak{k}$ respectively. Let $B(.,.)$ be the Killing form of $\mathfrak{g}$, $\mathfrak{g} = \mathfrak{k} \oplus \mathfrak{p}$ the Cartan decomposition, Θ the Cartan involution of $\mathfrak{g}$ (or of G) with respect to K and $(.,.)_\Theta$ the scalar product on $\mathfrak{g}$ defined by $(X_1,X_2) = -B(X_1,\Theta(X_2))$, $X_1,X_2 \in \mathfrak{g}$

Now let (P,S) be a split parabolic subgroup of G with split component A. For all details concerning parabolic subgroups the reader is referred to [66,Ch.2] , [78,I] . Among the split components of P there is a unique one which is Θ-stable. This is the special split component of (P,S) . Throughout this book we shall assume that the split component of (P,S) is Θ-stable. Let U be the unipotent radical of P and let L be the centralizer of A in G. L is a Levi subgroup of P and $P = UL$. Since A is Θ-stable the same is true for L. As usual, introduce the associated admissible closed reductive subgroup M of G such that $L = UM$ with $M \cap A = \{1\}$. Then

$$P = UAM$$

is the Langlands decomposition of P with respect to the split component A and $S = UM$. Let $\mathfrak{m}$, $\mathfrak{a}$ and $\mathfrak{u}$ be the Lie algebras of M, A and U, respectively. Since $U\backslash S = M$, we get a canonical homomorphism

$$\pi_{P|M} : S \longrightarrow M \tag{2.1}$$

The rank of (P,S) is the dimension of the split component A. Throughout this paper we only shall consider split parabolic subgroups (P,S) of rank one.

DEFINITION 2.2. Let (P,S) be a split parabolic subgroup of G of rank one and let $\Gamma \subset S$ be a discrete uniform subgroup without torsion. The manifold $Y = \Gamma\backslash\tilde{Y}$ is called a cusp of rank one associated to (P,S) and Γ.

EXAMPLE 2.3. Let $G = (SL(2,\mathbb{R}))^n$ and $K = (SO(2))^n$. Then $G/K = H^n$, where H is the upper half-plane. Let F be a totally real number field of degree n over $\mathbb{Q}$ and let $\mathbf{M}$ be a complete $\mathbb{Z}$-module of F, i.e. $\mathbf{M}$ is an additive subgroup of F which is free abelian of rank n. Denote by $U^+_{\mathbf{M}}$ the subgroup of those units ε of $\mathcal{O}_F$ (the ring of integers of F) which are totally positive and satisfy $\varepsilon\mathbf{M} = \mathbf{M}$. $U^+_{\mathbf{M}}$ is free abelian of rank $n-1$ [50,p.200]. Let $\mathbf{V} \subset U^+_{\mathbf{M}}$ be a subgroup of finite index and set

$$\Gamma_{\mathbf{M},\mathbf{V}} = \left\{ \begin{pmatrix} \varepsilon^{1/2} & \varepsilon^{-1/2}\mu \\ 0 & \varepsilon^{-1/2} \end{pmatrix} \;\middle|\; \varepsilon \in \mathbf{V} \;,\; \mu \in \mathbf{M} \right\}$$

$\Gamma_{\mathbf{M},\mathbf{V}}$ is a subgroup of $SL(2,F)$. Now observe that there are n different embeddings of F into the real numbers. These embeddings will be denoted by $x \in F \longmapsto x^{(j)} \in \mathbb{R}$, $j=1,\dots,n$. Using the different embeddings of F into $\mathbb{R}$ we get a map $SL(2,F) \longrightarrow (SL(2,\mathbb{R}))^n$ by sending $\begin{pmatrix} a & b \\ c & d \end{pmatrix} \in SL(2,F)$ to

$$\left\{ \begin{pmatrix} a^{(1)} & b^{(1)} \\ c^{(1)} & d^{(1)} \end{pmatrix}, \dots, \begin{pmatrix} a^{(n)} & b^{(n)} \\ c^{(n)} & d^{(n)} \end{pmatrix} \right\} \in (SL(2,\mathbb{R}))^n .$$

This is obviously an embedding. In particular, $\Gamma_{\mathbf{M},\mathbf{V}}$ can be considered as a subgroup of $(SL(2,\mathbb{R}))^n$. This subgroup is discrete. Let B be the subgroup of $SL(2,\mathbb{R})$ consisting of upper triangular matrices and set $P = B^n$. P is a parabolic subgroup of G. The unipotent radical U of P is given by

$$U = \left\{ \left(\begin{pmatrix} 1 & x_1 \\ 0 & 1 \end{pmatrix}, \dots, \begin{pmatrix} 1 & x_n \\ 0 & 1 \end{pmatrix} \right) \;\middle|\; x_1,\dots,x_n \in \mathbb{R} \right\} .$$

Let

$$M = \left\{ \left(\begin{pmatrix} \lambda_1 & 0 \\ 0 & \lambda_1^{-1} \end{pmatrix}, \dots, \begin{pmatrix} \lambda_n & 0 \\ 0 & \lambda_n^{-1} \end{pmatrix} \right) \;\middle|\; \lambda_i \in \mathbb{R}^\times \;,\; \lambda_1 \cdots \lambda_n = \pm 1 \right\}$$

and set $S = UM$. Then (P,S) is a split parabolic subgroup of G of rank one. The split component A is given by

$$A = \left\{ \left(\begin{pmatrix} \lambda & 0 \\ 0 & \lambda^{-1} \end{pmatrix}, \dots, \begin{pmatrix} \lambda & 0 \\ 0 & \lambda^{-1} \end{pmatrix} \right) \middle| \ \lambda \in \mathbb{R}^+ \right\}.$$

Since $\mathbf{V}$ is a group of units we have $\varepsilon^{(1)} \dots \varepsilon^{(n)} = 1$ for each $\varepsilon \in \mathbf{V}$. Thus $\Gamma_{\mathbf{M},\mathbf{V}}$ is contained in S. The complete $\mathbb{Z}$-module $\mathbf{M}$ is mapped injectively into $\mathbb{R}^n$ by sending $\mu \longrightarrow (\mu^{(1)},\dots,\mu^{(n)})$ and the image is a lattice in $\mathbb{R}^n$.

Moreover $\mathbf{V}$ acts on $\mathbf{M}$ in this representation by component-wise multiplication. This can be extended to an action of $\mathbf{V}$ on $\mathbb{R}^n$ by $\varepsilon.x = (\varepsilon^{(1)}x_1,\dots,\varepsilon^{(n)}x_n)$, $\varepsilon \in \mathbf{V}$, $x \in \mathbb{R}^n$. Let $\Gamma_M = M \cap (\Gamma_{\mathbf{M},\mathbf{V}} \cdot U)$. Then we have

$$\Gamma_M = \left\{ \left(\begin{pmatrix} (\varepsilon^{(1)})^{1/2} & 0 \\ 0 & (\varepsilon^{(1)})^{-1/2} \end{pmatrix}, \dots, \begin{pmatrix} (\varepsilon^{(n)})^{1/2} & 0 \\ 0 & (\varepsilon^{(n)})^{-1/2} \end{pmatrix} \right) \middle| \ \varepsilon \in \mathbf{V} \right\}$$

Sending $\varepsilon \in \mathbf{V}$ to $(\log \varepsilon^{(1)},\dots,\log \varepsilon^{(n)}) \in \mathbb{R}^n$ identifies $\mathbf{V}$ with a lattice in the hyperplane $\Sigma x_i = 0$ of $\mathbb{R}^n$. This shows that $\Gamma_M\backslash M$ is the disjoint union of 2^{n-1} tori of dimension $n-1$ and $\Gamma_{\mathbf{M},\mathbf{V}}\backslash S$ is a torus bundle over $\Gamma_M\backslash M$ with fibre $\Gamma_{\mathbf{M},\mathbf{V}} \cap U\backslash U = \mathbf{M}\backslash\mathbb{R}^n$. In particular, $\Gamma_{\mathbf{M},\mathbf{V}}\backslash S$ is compact. Thus $\Gamma_{\mathbf{M},\mathbf{V}}\backslash H^n$ is a cusp of rank one in the sense of Definition 2.2. More generally we can consider any group consisting of matrices $\begin{pmatrix} \varepsilon^{1/2} & \varepsilon^{1/2}\mu \\ 0 & \varepsilon^{-1/2} \end{pmatrix}$ with $\varepsilon \in \mathbf{V}$, $\mu \in F$, and $\mu \in \mathbf{M}$ if $\varepsilon = 1$, such that the sequence

$$0 \longrightarrow \mathbf{M} \longrightarrow \Gamma \longrightarrow \mathbf{V} \longrightarrow 1$$

is exact, where $\mathbf{M} \longrightarrow \Gamma$ is the map $\mu \longrightarrow \begin{pmatrix} 1 & \mu \\ 0 & 1 \end{pmatrix}$ and $\Gamma \longrightarrow \mathbf{V}$ is given by $\begin{pmatrix} \lambda & \mu \\ 0 & \lambda^{-1} \end{pmatrix} \longrightarrow \lambda^2$. Then $\Gamma\backslash H^n$ is a cusp of rank one. This example is of particular interest because it is related to Hirzebruch's conjecture.

Let (P,S) and Γ be as above and let $P = U.A.M$ be the Langlands decomposition of P with respect to the split component A of P. Put

$$\Gamma_M = M \cap (U \cdot \Gamma) = \pi_{P|M}(\Gamma) \quad . \qquad (2.4)$$

<u>LEMMA 2.5</u>. Γ_M is a discrete subgroup of M, $\Gamma_M\backslash M$ and $\Gamma \cap U\backslash U$ are compact and the sequence

$$1 \longrightarrow \Gamma \cap U \longrightarrow \Gamma \longrightarrow \Gamma_M \longrightarrow 1$$

is exact where the map $\Gamma \longrightarrow \Gamma_M$ is induced by the projection $\pi_{P|M}$. Finally, $\pi_{P|M}$ induces a fibration $\Gamma\backslash S \to \Gamma_M\backslash M$ with fibre $\Gamma \cap U\backslash U$.

PROOF. Consider the fibration $\Gamma\backslash S \to (U.\Gamma)\backslash S$. Its fibre is $\Gamma\backslash(U.\Gamma) = \Gamma\cap U\backslash U$. Since $M = U\backslash S$ and $\Gamma_M = U\backslash(U.\Gamma)$, one has $\Gamma_M\backslash M = (U\cdot\Gamma)\backslash S$. This shows that $\Gamma \cap U\backslash U$ and $\Gamma_M\backslash M$ are compact. Let $\{\delta_n\}_{n\in\mathbb{N}} \subset \Gamma_M$ be such that $\delta_n \to 1$ in M . There exists $\{u_n\}_{n\in\mathbb{N}} \subset U$ with $\gamma_n = \delta_n u_n \in \Gamma$. Since $\Gamma \cap U\backslash U$ is compact, we can assume that $\{u_n\}_{n\in\mathbb{N}}$ is contained in a compact set. Since Γ is discrete, this implies that $\delta_n = 1$ for $n \geq n_o$ and therefore $\delta_n = 1$ for $n \geq n_o$. The exactness of the sequence $1 \longrightarrow \Gamma \cap U \longrightarrow \Gamma \longrightarrow \Gamma_M \longrightarrow 1$ is clear by the remarks above. Q.E.D.

Let λ be the unique simple root of (P,A) with associated quasi-character $\alpha : A \to \mathbb{R}^+$. Let $g \in G$. Then g admits a decomposition $g = s.a_g k$ where $k \in K$, $a_g \in A$, $s \in S$. The factor a_g is unique. Let Λ_P be the function on G defined by

$$\Lambda_P(g) = \alpha(a_g) \ . \tag{2.6}$$

We shall often write $\Lambda(g)$ in place of $\Lambda_P(g)$ if it is clear which parabolic subgroup P we are considering.

Given $t \geq 0$, we set

$$A_t = \{a\in A \mid \alpha(a) > t\} \ . \tag{2.7}$$

For any compact neighborhood ω of 1 in S , the set

$$S_{t,\omega} = \omega.A_t.K$$

is called a Siegel domain in G (relative to (P,A)) . Let

$$K_M = M \cap K \ .$$

K_M is a maximal compact subgroup of M . Since M is θ-stable, we have $S \cap K = M \cap K$. By θ_M and B_M we denote the restriction to M of θ and B respectively. We put

$$X_M = M/K_M \ . \tag{2.8}$$

Since P is a parabolic subgroup of G , we have $P/P \cap K = G/K = \tilde{Y}$. But $P \cap K = S \cap K$ and $S \cap K = M \cap K$. Therefore we get an isomorphism

$$\tilde{\xi} : A \times U \times X_M \to \tilde{Y} \ . \tag{2.9}$$

Let $t \geq 0$. Then we define

$$\tilde{Y}_t = \tilde{\xi}(A_t \times U \times X_M) \tag{2.10}$$

Let $Z = S/S \cap K$. The homomorphism $\pi_{P|M} : S \longrightarrow M$ is an isomorphism of $S \cap K$ onto K_M and, by Lemma 2.5, $\pi_{P|M}$ induces a fibration

$$\Gamma\backslash Z \longrightarrow \Gamma_M\backslash X_M , \tag{2.11}$$

with fibre $\Gamma \cap U\backslash U$. From (2.9) we get an isomorphism

$$\xi : A \times \Gamma\backslash Z \longrightarrow \Gamma\backslash\tilde{Y} = Y . \tag{2.12}$$

For $t \geq 0$ we set

$$Y_t = \xi(A_t \times \Gamma\backslash Z) . \tag{2.13}$$

We observe that $\tilde{Y}_t$ defined by (2.10) is invariant under Γ and $Y_t = \Gamma\backslash \tilde{Y}_t$. Each manifold of this type will also be called a cusp of rank one.

Associated to the parabolic subgroup P there is a decomposition of the metric on the symmetric space $\tilde{Y} = G/K$ (c.f. [19, §4]) . Let $\mathfrak{u}$ be the Lie algebra of U . Let $m(\lambda)$ be the multiplicity of λ and $m(2\lambda)$ the multiplicity of 2λ . Then, in an obvious notation, $\mathfrak{u}$ admits the direct sum decomposition

$$\mathfrak{u} = \mathfrak{u}_\lambda \oplus \mathfrak{u}_{2\lambda}$$

where $\dim \mathfrak{u}_\lambda = m(\lambda)$ and $\dim \mathfrak{u}_{2\lambda} = m(2\lambda)$. Let h_λ (resp. $h_{2\lambda}$) be the right invariant scalar product on U which is zero on $\mathfrak{u}_{2\lambda}$ (resp. $\mathfrak{u}_\lambda$) and equal to $(.,.)_\theta$ on $\mathfrak{u}_\lambda$ (resp. $\mathfrak{u}_{2\lambda}$) . For $x = m.K_M$, $m \in M^\circ$, and $\alpha \in \{\lambda, 2\lambda\}$ set $du_\lambda^2(x) = \frac{1}{2}(\mathrm{Int}\ m)^* h_\alpha$. Then one has

$$ds^2 = da^2 + dx^2 + a^{-2\lambda} du_\lambda^2(x) + a^{-4\lambda} du_{2\lambda}^2(x)$$

where dx^2 (resp. da^2) is the invariant metric on X_M (resp. A) induced by the restriction of the Killing form [19, Prop. 4.3]. This is also locally the description of the metric on the cusp. The volume element is given by

$$d\mathrm{vol} = a^{-2\rho} da\, d\mathrm{vol}(x)\, d\mathrm{vol}(u)$$

where $2\rho = m(\lambda)\lambda + m(2\lambda)2\lambda$. Note that $d\mathrm{vol}(u)$ is independent of x . We shall identify A with $\mathbb{R}^+$ by sending $a \in A$ to $\exp(\lambda(\log a)/|\lambda|)$. Then the metric (2.14) is given by

$$ds^2 = \frac{dr^2}{r^2} + dx^2 + r^{-2|\lambda|}du^2(x) + r^{-4|\lambda|}du^2_{2\lambda}(x) \tag{2.16}$$

and the volume element equals

$$dvol = \frac{1}{|\lambda|} r^{-(m+1)} dr\, dvol(x) dvol(u) \tag{2.17}$$

where $m = m(\lambda)|\lambda| + 2m(2\lambda)|\lambda|$.

We continue by recalling some facts concerning locally invariant differential operators (see [15],[61] for more details). Given a unitary representation σ of K on a finite dimensional complex vector space V , we denote by $\tilde{V}$ the associated homogeneous vector bundle over $\tilde{Y}$. The space $C^\infty(\tilde{Y},\tilde{V})$ (resp. $C_c(\tilde{Y},\tilde{V})$) of all C^∞-sections (resp. C^∞-sections with compact supports) of $\tilde{V}$ will be identified with the K-invariant subspace $(C^\infty(G) \otimes V)^K$ (resp. $(C_c^\infty(G) \otimes V)^K$) of $C^\infty(G) \otimes V$ (resp. $C_c^\infty(G) \otimes V$) with respect to the action $k \longmapsto R(k) \otimes \sigma(k)$ of K , R being the right regular representation of G . Similarly, the space of L^2-sections of $\tilde{V}$ will be identified with $(L^2(G) \otimes V)^K$. The homogeneous vector bundle $\tilde{V}$ over $\tilde{Y}$ can be pushed down to a "locally homogeneous" bundle $V = \Gamma\backslash\tilde{V}$ over $Y = \Gamma\backslash\tilde{Y}$. As above, its space of C^∞-sections $C^\infty(Y,V)$ will be identified with the K-invariant part $(C^\infty(\Gamma\backslash G) \otimes V)^K$ of $C^\infty(\Gamma\backslash G) \otimes V$ with respect to the representation $R_{\Gamma\backslash G} \otimes \sigma$, where $R_{\Gamma\backslash G}$ stands for the right regular representation of G on $C^\infty(\Gamma\backslash G)$. The same applies to the spaces $C_c^\infty(Y,V)$ and $L^2(Y,V)$ of C^∞-sections with compact supports and L^2-sections of V respectively. Sometimes we shall identify sections of V and maps $\varphi: \Gamma\backslash G \longrightarrow V$ satisfying $\varphi(gk) = \sigma(k)^{-1}\varphi(g)$ for all $k \in K$, $g \in G$.

Let $\varphi : \Gamma\backslash G \rightarrow V$ be a vector valued locally bounded measurable function. Consider φ as a Γ-invariant function $\varphi : G \rightarrow V$. Its constant term is defined by

$$\varphi_o(g) = \int_{\Gamma\cap U\backslash U} \varphi(ug)du \quad , \tag{2.18}$$

where we assume that the measure on U is normalized by the condition $vol(\Gamma \cap U\backslash U) = 1$. Since U is a normal subgroup of P , φ_o is clearly left invariant under $\Gamma\cdot U$. Thus φ_o can be considered as a function $\varphi_o : (\Gamma\cdot U)\backslash G \rightarrow V$. $\varphi \in L^2(\Gamma\backslash G)$ is called a cusp form if $\varphi_o(g) = 0$ for a.e. $g \in G$. The space of cusp forms on $\Gamma\backslash G$ will be denoted by $L_o^2(\Gamma\backslash G)$. We set

$$L_o^2(Y,V) = (L_o^2(\Gamma\backslash G) \otimes V)^K \quad . \tag{2.19}$$

This is the space of cusp forms on $\Gamma\backslash G$ of type σ .

Let $\mathfrak{g}_{\mathbb{C}}$ be the complexification of $\mathfrak{g}$ and $\mathfrak{U}(\mathfrak{g}_{\mathbb{C}})$ the universal enveloping algebra of $\mathfrak{g}_{\mathbb{C}}$. By $\mathfrak{Z}(\mathfrak{g}_{\mathbb{C}})$ we denote the center of $\mathfrak{U}(\mathfrak{g}_{\mathbb{C}})$. Let σ_1, σ_2 be two finite-dimensional unitary representations of K on V_1 and V_2 , respectively, and denote by $(\mathfrak{U}(\mathfrak{g}_{\mathbb{C}}) \otimes \mathrm{Hom}(V_1,V_2))^K$ the subspace of all elements in $\mathfrak{U}(\mathfrak{g}_{\mathbb{C}}) \otimes \mathrm{Hom}(V_1,V_2)$ which are invariant by the representation $k \longmapsto \mathrm{Ad}(k) \otimes (\sigma_1(k^{-1}) \otimes \sigma_2(k))$ of K. Given

$$D = \sum_i X_i \otimes A_i \in (\mathfrak{U}(\mathfrak{g}_{\mathbb{C}}) \otimes \mathrm{Hom}(V_1,V_2))^K$$

we put

$$\tilde{\mathcal{D}} = \sum_i R(X_i) \otimes A_i \ . \tag{2.20}$$

This formula defines a differential operator $\tilde{\mathcal{D}} : C_c^\infty(\tilde{V}_1) \longrightarrow C_c^\infty(\tilde{V}_2)$ which is G-invariant. Conversely, any G-invariant differential operator from $C_c^\infty(\tilde{V}_1)$ to $C_c^\infty(\tilde{V}_2)$ arises in this manner. The formal adjoint $\tilde{\mathcal{D}}^*$ corresponds to

$$D^* = \sum_i X_i^* \otimes A_i^*$$

where A_i^* is the adjoint map of $A_i : V_1 \longrightarrow V_2$ and X_i^* is the image of X_i under the canonical anti-involution of $\mathfrak{U}(\mathfrak{g}_{\mathbb{C}})$ which sends $X+\sqrt{-1}Y \in \mathfrak{g}_{\mathbb{C}}$ into $-X+\sqrt{-1}Y$, $X,Y \in \mathfrak{g}$. Since $\tilde{\mathcal{D}}$ is G-invariant, it defines a differential operator

$$\mathcal{D} : C_c^\infty(Y,\mathcal{V}_1) \longrightarrow C_c^\infty(Y,\mathcal{V}_2) \ . \tag{2.21}$$

This is the locally invariant differential operator associated to D. D is called elliptic if $\tilde{\mathcal{D}}$ is so.

Finally, we recall the definition of Harish-Chandra's L^p-Schwartz spaces $\mathcal{C}^p(G)$, $p > 0$, of p-integrable rapidly decreasing functions. For $g \in G$, let $\delta(g)$ be the geodesic distance between the cosets K and gK in the symmetric space G/K and let Ξ be the spherical function on G which is given by

$$\Xi(g) = \int_K e^{-\rho(H(kg))} dk$$

with the usual notation. Consider a finite-dimensional Hilbert space V. If $\Phi : G \longrightarrow V$ is a C^∞-function and $D_1, D_2 \in \mathfrak{U}(\mathfrak{g}_{\mathbb{C}})$, assign the usual meaning to $\Phi(D_1;g;D_2)$ [78,II,p.104] . Let $p > 0$. Then $\mathcal{C}^p(G,V)$ is the space of functions $\Phi \in C^\infty(G,V)$ such that

$$\sup_{g\in G}((1+\delta(g))^r \,\Xi(g)^{-2/p}\,\|\,\Phi(D_1;g;D_2)\,\|) < \infty \quad ,$$
$$\text{for all} \quad r\in \mathbb{R}^+ \,,\ D_1,D_2\in \mathfrak{U}(\mathfrak{g}_{\mathbb{C}}) \ . \tag{2.22}$$

For $V = \mathbb{C}$ set $C^p(G) = C^p(G,V)$. If σ is a finite-dimensional unitary representation of K on a complex vector space V, we set

$$C^p(G,\sigma) = \{\ \Phi\in C^p(G,\mathrm{End}(V))\mid \Phi(k_1gk_2) = \sigma(k_1)\circ\Phi(g)\circ\sigma(k_2)$$
$$\text{for all} \quad g\in G \ ,\ k_1,k_2\in K \quad \} \quad . \tag{2.23}$$

CHAPTER III

THE HEAT EQUATION ON THE CUSP

In this chapter we shall construct the heat kernel for locally invariant operators on a cusp of rank one which, up to a zero-order operator, are given by the action of the Casimir element of $\mathfrak{g}$. In §7, we shall use this kernel as a parametrix at infinity for the heat kernel of certain operators which are locally invariant at infinity.

Let $Y = \Gamma\backslash G/K$ be a cusp of rank one with respect to the split parabolic subgroup (P,S) and let $\mathcal{V}$ be a locally homogeneous Hermitian vector bundle over Y defined by a unitary representation μ of K on the complex vector space V. If $\Omega \in \mathfrak{Z}(\mathfrak{g}_{\mathbb{C}})$ is the Casimir element of $\mathfrak{g}$, then $-\Omega \otimes \mathrm{Id}_V \in \mathfrak{U}(\mathfrak{g}_{\mathbb{C}}) \otimes \mathrm{End}(V)$ is K-invariant. Let $L \in \mathrm{End}_K(V)$ be such that $L = L^*$ and consider the locally invariant differential operator

$$\Delta_\mu : C_c^\infty(Y,\mathcal{V}) \longrightarrow C_c^\infty(Y,\mathcal{V}) \tag{3.1}$$

defined by the element $-\Omega \otimes \mathrm{Id}_V + \mathrm{Id} \otimes L \in (\mathfrak{U}(\mathfrak{g}_{\mathbb{C}}) \otimes \mathrm{End}(V))^K$. Since $\Omega^* = \Omega$ and $L^* = L$, Δ_μ is formally selfadjoint. If R_Γ is the right regular representation of G on $C_c^\infty(\Gamma\backslash G)$, then Δ_μ is the restriction of $-R_\Gamma(\Omega) \otimes \mathrm{Id}_V + \mathrm{Id} \otimes L$ to the K-invariant part $(C_c^\infty(\Gamma\backslash G) \otimes V)^K$ of $C_c^\infty(\Gamma\backslash G) \otimes V$.

Let $\tilde{\nabla}$ be the canonical invariant connection of $\tilde{\mathcal{V}}$ and $\tilde{\Delta}_\mu$ the lift of Δ_μ to a G-invariant operator on $C_c^\infty(\tilde{Y},\tilde{\mathcal{V}})$. Let

$$Q = \int_K R(k) \otimes \mu(k)\,dk\,. \tag{3.2}$$

Q is the orthogonal projection of $L^2(G) \otimes V$ onto $(L^2(G) \otimes V)^K$ and one has $\tilde{\Delta}_\mu = -Q(R(\Omega) \otimes \mathrm{Id}_V)Q + Q(\mathrm{Id} \otimes L)Q$. Let $X_1,\ldots,X_p$ be an orthonormal basis for $\mathfrak{p}$ with respect to $B(.,.)$ and let $Y_1,\ldots,Y_m$ be an orthonormal basis for $\mathfrak{k}$ with respect to $-B(.,.)$. Set $\Omega_K = -\sum_{j=1}^m Y_j^2$. Recall that $\Omega = \sum_{i=1}^p X_i^2 + \Omega_K$ and $\Omega_K \in \mathfrak{Z}(\mathfrak{k}_{\mathbb{C}})$, where $\mathfrak{Z}(\mathfrak{k}_{\mathbb{C}})$ denotes the center of the universal enveloping algebra of $\mathfrak{k}_{\mathbb{C}}$. Moreover, the connection Laplacian $\tilde{\nabla}^*\tilde{\nabla}$ is given by

$\tilde{\nabla}^*\tilde{\nabla} = -Q(\sum_{i=1}^{p} R(X_i)^2)Q$. Therefore we get

$$\tilde{\Delta}_\mu = -Q(R(\Omega) \otimes Id_V)Q + Q(Id \otimes L)Q = \tilde{\nabla}^*\tilde{\nabla} + Q(Id \otimes L_1)Q \quad (3.3)$$

where $L_1 \in End_K(V)$ is the element $-2\mu(\Omega_K)+L$. This shows that $\tilde{\Delta}_\mu$ is an elliptic differential operator on $C_c^\infty(\tilde{Y},\tilde{V})$. It follows from [61, Corollary 1.2] that the closure of Δ_μ in $L^2(Y,V)$, for which we shall use the same notation, is a selfadjoint operator. This is the unique selfadjoint extension of Δ_μ to an unbounded operator in the Hilbert space $L^2(Y,V)$. Moreover, by (3.3), there exists a constant $c \geq 0$ such that $\Delta_\mu + cId \geq 0$. Therefore the heat semigroup $\exp(-t\Delta_\mu)$, $t \geq 0$, exists. In this section we shall study the kernel of the heat operator $\exp(-t\Delta_\mu)$. For this purpose we need some preparatory material which we shall discuss now.

Let V be a finite-dimensional Hilbert space and let $\mathcal{C}^p(G,V)$, $p > 0$, be Harish-Chandra's L^p-Schwartz space as defined in §2.

LEMMA 3.4. Let $\phi \in \mathcal{C}^1(G,V)$. Then there exists a compact symmetric neighborhood $\mathcal{U}$ of the identity in G and a nonnegative integrable function ϕ_o on G such that for all $D_1, D_2 \in \mathfrak{U}(\mathfrak{g}_\mathbb{C})$ and all $x \in G$ one has

$$\|\phi(D_1;x;D_2)\| \leq C \int_{\mathcal{U}} \phi_o(yx)\,dy \quad ,$$

where the constant C depends only on $D_1, D_2 \in \mathfrak{U}(\mathfrak{g}_\mathbb{C})$.

PROOF. Let $\mathcal{U}$ be any compact symmetric neighborhood of the identity in G . Let Ξ and δ be the functions introduced at the end of §2. By [78, II, Prop. 8.3.7.5] there exists $r \geq 0$ such that

$$\int_G \Xi^2(x)(1+\delta(x))^{-r}dx < \infty \quad .$$

Set $\phi_o(x) = \Xi^2(x)(1+\delta(x))^{-r}$. If we use Corollary 8.1.2.2 and Proposition 8.3.7.2 in [78,II] , it follows that there exists a constant C_1 such that

$$\phi_o(x) \leq C_1 \int_{\mathcal{U}} \phi_o(yx)\,dy \ , \qquad x \in G \ .$$

Let $D_1, D_2 \in \mathfrak{U}(\mathfrak{g}_\mathbb{C})$. Since $\phi \in \mathcal{C}^1(G,V)$, there exists another constant C_2 such that

$$\|\phi(D_1;x;D_2)\| \leq C_2\, \phi_o(x) \ .$$

Finally, note that ϕ_o is nonnegative and integrable. Q.E.D.
For any element $T \in \mathrm{End}(\mathfrak{g})$ let T^* denote its adjoint relative to $(.,.)_\theta$. On G we consider the standard semi-norm, which is defined by

$$\|g\|^2 = \mathrm{tr}(\mathrm{Ad}(g)\mathrm{Ad}(g)^*) , \quad g \in G . \tag{3.5}$$

Let $n = \dim \mathfrak{g}$. If $x \in GL(\mathfrak{g})$, then we have $(\mathrm{tr}(xx^t))^n \geq (\det(x))^2$. Since $\det \mathrm{Ad}(g) = \pm 1$ for all $g \in G$, the function (3.5) satisfies $\|g\| \geq 1$ for all $g \in G$. Moreover, one has $\|g_1g_2\| \leq \|g_1\| . \|g_2\|$, $g_1, g_2 \in G$ and $\|g\| = \|g^{-1}\|$, $g \in G$ [78, II , 8.1.5.1] .

<u>LEMMA 3.6</u>. Let $\phi \in C^1(G,V)$. Then there exists $r > 0$ such that for all $D_1, D_2 \in \mathfrak{U}(\mathfrak{g}_{\mathbb{C}})$ and $x, y \in G$ one has

$$\sum_{\gamma\in\Gamma} \|\phi(D_1;x^{-1}\gamma y;D_2)\| \leq C\|y\|^r .$$

The constant C depends on D_1, D_2 and ϕ.

<u>PROOF</u>. Choose a compact symmetric neighborhood U of the identity in G and a nonnegative integrable function ϕ_o on G as in Lemma 3.4. Then, for all $D_1, D_2 \in \mathfrak{U}(\mathfrak{g}_{\mathbb{C}})$, there exists a constant C such that

$$\|\phi(D_1;x^{-1}\gamma y;D_2)\| \leq C \int_U \phi_o(zx^{-1}\gamma y)\,dz , \tag{3.7}$$

for all $x, y \in G$. It is clear that $Ux^{-1}\gamma_1 y \cap Ux^{-1}\gamma_2 y \neq \emptyset$ if $\gamma_1\gamma_2^{-1} \in$ $\in xU^{-1}Ux^{-1}$. For $t > 0$ let $G[t]$ be the set of all elements $y \in G$ such that $\|y\| \leq t$. Using the properties of the function (3.5) stated above, it follows that the set $xU^{-1}Ux^{-1}$ is contained in $G[b\|x\|^2]$ for a certain constant b. Thus

$$\#(\Gamma \cap xU^{-1}Ux^{-1}) \leq \#(\Gamma \cap G[b\|x\|^2]) . \tag{3.8}$$

Since Γ is a discrete subgroup of G, there exists a constant C_1 such that $\#(\Gamma \cap G[t]) \leq C_1 \mathrm{vol}(G[t])$. Let Z_G be the center of G and put $G' = G/Z_G$. G' is a connected semi-simple Lie group with trivial center. Therefore, the adjoint representation $\mathrm{Ad} : G' \to GL(\mathfrak{g})$ is injective and we can apply Lemma 37 of [41] to conclude that $\mathrm{vol}(G'[t]) \leq C_2t^m$, $t > 0$, for certain constants C_2 and m. Now recall that, by assumption, Z_G is finite. Therefore $G \to G'$ is a finite covering and this implies $\mathrm{vol}(G[t]) \leq C_3t^m$. Hence, by (3.8), there exist constants C_4 and r such that

$$\#(\Gamma \cap xU^{-1}Ux^{-1}) \leq C_4 \|x\|^r .$$

By (3.7) we can conclude that

$$\sum_{\gamma\in\Gamma} \|\phi(D_1;x^{-1}\gamma y;D_2)\| \le C_5 \|x\|^r \int_G \phi_o(z)\,dz \, . \quad \text{Q.E.D.}$$

COROLLARY 3.9. Let $\phi \in C^1(G,V)$. Then the series

$$K(x,y) = \sum_{\gamma\in\Gamma} \phi(x^{-1}\gamma y)$$

is absolutely convergent for all x , $y \in G$. The convergence is uniform on compacta and $K(x,y)$ is C^∞ in both x and y .

We now turn to the construction of the kernel of the heat operator $\exp(-t\Delta_\mu)$. This kernel is obtained by averaging over Γ the kernel of the heat operator $\exp(-t\tilde{\Delta}_\mu)$ where $\tilde{\Delta}_\mu$ is the lift of Δ_μ to $\tilde{Y}$. Thus the first step is to study the kernel of the heat operator $\exp(-t\tilde{\Delta}_\mu)$. Consider the G-invariant differential operator

$$\tilde{\Delta}_\mu : C_c^\infty(\tilde{Y},\tilde{V}) \to C_c^\infty(\tilde{Y},\tilde{V})$$

which is associated to the element $-\Omega \otimes \mathrm{Id} + \mathrm{Id} \otimes L \in (\mathcal{U}(\mathfrak{g}_{\mathbb{C}}) \otimes \mathrm{End}(V))^K$. As we have seen above, $\tilde{\Delta}_\mu$ is a formally selfadjoint elliptic differential operator on $C_c^\infty(\tilde{Y},\tilde{V})$. Appealing again to Corollary 1.2 in [61], it follows that the closure of $\tilde{\Delta}_\mu$ in L^2 , which we will continue to denote $\tilde{\Delta}_\mu$, is a selfadjoint operator. Moreover, by (3.3), this operator is bounded from below and we can form the heat semi-group $\exp(-t\tilde{\Delta}_\mu)$ $t \geq 0$. For each $t > 0$, the bounded operator

$$\exp(-t\tilde{\Delta}_\mu) : L^2(\tilde{Y},\tilde{V}) \to L^2(\tilde{Y},\tilde{V})$$

is a smoothing operator which commutes with the representation of G on $L^2(\tilde{Y},\tilde{V})$. Therefore, for each $t > 0$, $\exp(-t\tilde{\Delta}_\mu)$ is an integral operator defined by a smooth G-invariant kernel. Being G-invariant it corresponds to a function $h_t : G \to \mathrm{End}(V)$ which is C^∞ , square integrable and satisfies the covariance property

$$h_t(k_1 g k_2) = \mu(k_1).h_t(g).\mu(k_2) \, , \quad g \in G \, , \quad k_1,k_2 \in K \, , \tag{3.10}$$

(see [60]) . The function h_t is the kernel of the heat operator $\exp(-t\tilde{\Delta}_\mu)$, i.e.

$$(\exp(-t\tilde{\Delta}_\mu)\varphi)(g) = \int_G h_t(g^{-1}g')\,\varphi(g')\,dg' \tag{3.11}$$

for $\varphi \in L^2(\tilde{Y},\tilde{V})$, $g \in G$.

In order to get more information about the heat kernel

h_t we observe that h_t is closely related to the standard heat kernel on G . This is the approach used by Barbasch and Moscovici [15,§2] in the case of the spinor Laplacians. The results of [15,§2] can be easily extended to our case. For the sake of completeness we give some details. Let

$$\Delta = -\Omega+2\Omega_K \in \mathcal{U}(\mathfrak{g}_{\mathbb{C}}) , \tag{3.12}$$

where Ω_K is the Casimir element of $\mathfrak{k}$. Then $R(\Delta)$ is the Laplacian operator on G with respect to the left invariant Riemannian metric on G induced by the scalar product $(.,.)_\Theta$ on $\mathfrak{g}$. Consider the heat semigroup $\exp(-tR(\Delta))$ generated by $R(\Delta)$.$\exp(-tR(\Delta))$ is a G-invariant smoothing operator acting on $L^2(G)$. Therefore, as above, we can conclude tat there exists a function $p_t \in L^2(G) \cap C^\infty(G)$ such that

$$(\exp(-tR(\Delta))f)(g) = \int_G p_t(g^{-1}g')f(g')\, dg' , \tag{3.13}$$

$f \in L^2$, $g \in G$. Moreover one has $p_t \in L^1(G)$ [64, Sect. 8] , so that (3.13) can be written as

$$\exp(-tR(\Delta)) = R(p_t) ,$$

R being the right regular representation of G on $L^2(G)$. Let Q be the operator defined by (3.2). Using the same calculations as in [15, p. 160] , it is easy to see that

$$\exp(-t\tilde{\Delta}_\mu) = Q(e^{-tR(\tilde{\Delta})} \otimes e^{t(2\mu(\Omega_K)-L)})Q \tag{3.14}$$

and this implies that

$$h_t(g) = \int_K \int_K p_t(k_1gk_2)\mu(k_1)^{-1}\cdot e^{t(2\mu(\Omega_K)-L)}\cdot\mu(k_2)^{-1}dk_1dk_2 . \tag{3.15}$$

Let $\mathcal{C}^p(G,\mu)$, $p > 0$, be Harish-Chandra's L^p-Schwartz space of type μ defined by (2.23). Using (3.15) we may deduce

<u>PROPOSITION 3.16</u>. Let $t > 0$. Then $h_t \in \mathcal{C}^p(G,\mu)$ for all $p > 0$.

The proof is the same as the proof of Proposition 2.4 in [15]. Therefore it will be omitted.

If we identify $(L^2(G) \otimes V)^K$ and $L^2(\tilde{Y},\tilde{\mathcal{V}})$ then $h_t(g^{-1}g')$ corresponds to a section $E(x,y,t)$ of $\tilde{\mathcal{V}} \boxtimes \tilde{\mathcal{V}}^*$ over $\tilde{Y} \times \tilde{Y} \times \mathbb{R}^+$. To obtain an estimate on the behavior of this kernel as $t \to 0$ we use the construction of Donnelly [26] for the fundamental solution of the heat equation. First observe that, in view of (3.3), the principal symbol of $\tilde{\Delta}_\mu$ is given by $\sigma(\tilde{\Delta}_\mu)(x,\xi) = \langle\xi,\xi\rangle_x I$ where $\langle.,.\rangle_x$ is the Riemannian metric in $T^*_x\tilde{Y}$. Moreover, according to Borel [20, Theorem B] there exists a torsion free discrete subgroup $\Gamma \subset G$ with compact quotient $\Gamma\backslash G/K$. Therefore we can extend the method of [26] to construct the fundamental solution $E(x,y,t)$ of the heat equation for $\tilde{\Delta}_\mu$. There exists a unique fundamental solution $E(x,y,t)$ which satisfies the same properties as in [26, p. 488]. Moreover, $E(x,y,t)$ is symmetric and satisfies the semigroup property. We recall only property P4 : For each $T > 0$ there exists a constant $C > 0$ such that for $0 < t \leq T$ one has

$$\|(\frac{\partial}{\partial t})^i \nabla_x^j \nabla_y^k E(x,y,t)\| \leq Ct^{-n/2-i-j-k}\exp(-\frac{d^2(x,y)}{4t}) \tag{3.17}$$

where $n = \dim \tilde{Y}$, $i,j,k \in \mathbb{N}$ and $d(x,y)$ denotes the geodesic distance of $x,y \in \tilde{Y}$.

Now we claim that for each $t > 0$ and $y \in \tilde{Y}$, the function $x \to \exp(-d^2(x,y)/4t)$ is in $L^1(\tilde{Y}) \cap L^2(\tilde{Y})$. This is an immediate consequence of Lemma 4.4.5.9 in [78,I] (compare [78,I,p. 285]). In view of this observation, the integral

$$(P_t\varphi)(x) = \int_Y E(x,y,t)\varphi(y)dy , \quad \varphi \in L^2(\tilde{Y},\tilde{V}) ,$$

is well defined. Actually, P_t is a bounded operator on $L^2(\tilde{Y},\tilde{V})$. Using properties P1-P4 in [26, p. 488] it follows that $P_t\varphi \in L^2(\tilde{Y},\tilde{V}) \cap C^\infty(\tilde{Y},\tilde{V})$ for each $\varphi \in L^2(\tilde{V},\tilde{V})$ and one has $(\frac{\partial}{\partial t} + \tilde{\Delta}_\mu)P_t\varphi = 0$. Moreover, $P_t\varphi \longrightarrow \varphi$ in $L^2(\tilde{Y},\tilde{V})$ as $t \to 0$. Now we observe that a solution of the heat equation $(\frac{\partial}{\partial t} + \tilde{\Delta}_\mu)u = 0$ is uniquely determined by its initial data (see [52, IX, §13]). Therefore we have $P_t = \exp(-t\tilde{\Delta}_\mu)$, and this implies that the kernel $\tilde{E}(x,y,t)$ given by h_t coincides with $E(x,y,t)$.

Now, we can return to our original problem and investigate the kernel of the heat operator $\exp(-t\Delta_\mu)$ acting on $L^2(Y,V)$. Using Proposition 3.16 together with Corollary 3.9 it follows that the series

$$e(g,g',t) = \sum_{\gamma\in\Gamma} h_t(g^{-1}\gamma g') \tag{3.18}$$

is absolutely convergent for all $g,g' \in G$. The convergence is uniform on compacta and $e(g,g',t)$ is C^∞ in g,g' and t. Moreover, this kernel satisfies

$$e(\gamma_1 g,\gamma_2 g',t) = e(g,g',t) \qquad , \gamma_1,\gamma_2 \in \Gamma \tag{3.19}$$

and (3.10) implies that

$$e(gk_1,g'k_2,t) = \mu(k_1)^{-1} \circ e(g,g',t) \circ \mu(k_2) , \quad k_1,k_2 \in K .$$

Thus we can regard e as C^∞-function on $\Gamma\backslash G \times \Gamma\backslash G \times \mathbb{R}^+$ with values in $End(V)$ which satisfies the covariance property (3.19). Therefore we may identify e with a smooth section of $V \boxtimes V^*$, the external tensor product of V and V^* over $Y \times Y$. It satisfies

$$(\frac{\partial}{\partial t} + \Delta_\mu)e(z,z',t) = 0 ,$$

where Δ_μ acts either in the first or in the second variable. Using (3.17) one can establish a corresponding estimate on the behaviour of e as $t \to 0$. First we prove an auxiliary lemma. For $c > 0$ and $\omega \subset Z$ we set $S_{c,\omega} = \tilde{\xi}(A_c\omega)$ where $\tilde{\xi}$ is the isomorphism (2.9). Moreover, recall that, after identifying $\mathbb{R}^+$ and A, we have $\tilde{Y} = \mathbb{R}^+ \times Z$. For each $z \in \tilde{Y}$ we shall denote by $r(z) \in \mathbb{R}^+$ the first component of z with respect to this isomorphism. Then we have

LEMMA 3.20. Let $T,c > 0$ and $\omega \subset Z$, ω compact, be given. There exists a constant $C_1 > 0$ such that

$$\sum_{\gamma\in\Gamma} \exp(-\frac{d^2(z,\gamma z')}{4t}) \leq C_1(r(z)r(z'))^{m/2} \exp(-\frac{d^2(z,z')}{8t})$$

for $0 < t \leq T$ and $z,z' \in S_{c,\omega}$.

PROOF. Let $z,z' \in \tilde{Y}$. For $R > 0$ let $N(R)$ denote the number of elements $\gamma \in \Gamma$ such that $d(z,\gamma z') \leq R$. First, we shall estimate this number. For $x \in Z$ we denote by $B_\varepsilon(x)$ the ball of radius $\varepsilon > 0$ around x in Z. Since $\Gamma\backslash Z$ is compact there exists $\varepsilon > 0$ such that, for each $x \in Z$, one has $\gamma B_\varepsilon(x) \cap B_\varepsilon(x) = \emptyset$, for $\gamma \in \Gamma - \{1\}$. Assume that $z = (r,x)$, $z' = (r',x')$ with $x,x' \in Z$. Set $B(z') = [r'/2,2r'] \times B_\varepsilon(x')$. By $B(z,R)$ we shall denote the ball of radius R around z in $\tilde{Y}$. There exist constants $C_2,C_3 > 0$ such that $Vol(B(z,R)) \leqq C_2\exp(C_3R)$, uniformly for $z \in \tilde{Y}$ [26, Lemma 4.1]. Assume that $R > diam(B(z'))$. Then $z' \in B(z,R)$ implies $\gamma B(z') \subset B(z,2R)$ Therefore, $N(R)$ can be estimated by $Vol(B(z,R))/Vol(B(z')) \leqq$ $\leqq C_2\exp(C_3R)/Vol(B(z'))$. If we use (2.14), it follows that there exists a constant $C_4 > 0$ such that $Vol(B(z')) = C_4 Vol(B\ (x'))r'^m$, for $z' \in \tilde{Y}$. Thus we get

$$N(R) \leqq C_5 r'^m \exp(C_3R) \quad , \quad R > diam(B(z')) \ , \ z,z' \in \tilde{Y} .$$

Let $R > diam(B(z'))$. Then, using this estimation, we obtain

$$\sum_{\gamma\in\Gamma} \exp(-\frac{d^2(z,\gamma z')}{4t}) = \sum_{n=o}^{\infty} \sum_{\substack{\gamma\in\Gamma \\ Rn<d(z,\gamma z')\leq R(n+1)}} \exp(-\frac{d^2(z,\gamma z')}{4t}) \leq$$

$$\leq C_5\, r(z')^m \sum_{n=0}^{\infty} \exp(C_3R(n+1) - \frac{R^2n^2}{8t} - \frac{d^2(z,z')}{8t}) \quad .$$

Now observe that

$$\sum_{n=0}^{\infty} \exp(- \frac{R^2n^2}{8t} + C_3Rn) \leq 1 + \int_0^{\infty} \exp(- \frac{R^2x^2}{8t} + C_3Rx)\,dx \leq 1+C_6R^{-1}e^{C_7t} \quad .$$

Thus there exists a constant $C_8 > 0$ such that

$$\sum_{\gamma\in\Gamma} \exp(- \frac{d^2(z,\gamma z')}{4t}) \leq C_8e^{C_3R}\, r'^m \exp(- \frac{d^2(z,z')}{8t}) \tag{3.21}$$

for $0 < t \leq T$, $z,z' \in \tilde{Y}$ and $R > \max(\operatorname{diam}(B(z')),1)$. Now observe that there exists a constant $C > 0$ such that $\operatorname{diam}(B(z')) \leq C$ for $z' \in S_{c,\omega}$. If we use (3.21) and the corresponding inequality with the role of z and z' switched we get the desired result. Q.E.D.

We identify $z,z' \in Y$ with points in the universal cover $\tilde{Y}$ which realize the geodesic distance from z to z' in Y . By (3.17) and Lemma 3.20 we obtain

$$\begin{aligned} \|e(z,z',t)\| &\leq Ct^{-n/2} \sum_{\gamma\in\Gamma} \exp(- \frac{d^2(z,\gamma z')}{4t}) \leq \\ &\leq C_1\, t^{-n/2}(r(z)r(z'))^{m/2}\exp(- \frac{d^2(z,z')}{8t}) \ , \end{aligned} \tag{3.22}$$

uniformly for $0 < t \leq T$, $z,z' \in Y_c$, $c > 0$. The constant C_1 depends on T and c . The estimation (3.22) shows that $e(z,z',t)$ is a fundamental solution of the heat equation for Δ_μ . Moreover, using (3.21) it is easy to see that e is the kernel of the heat operator $\exp(-t\Delta_\mu)$. For this purpose one has to estimate $\operatorname{diam}(B(z))$ as $r(z) \to 0$.

Let $\omega_1 \subset U_\lambda$, $\omega_2 \subset U_{2\lambda}$ and $\omega_3 \subset X_M$ be compact subsets so that $\omega_1 \times \omega_2 \times \omega_3$ contains a fundamental domain of Γ acting on Z . Let $x_o \in X_M$ be fixed . Then, using (2.16) it is easy to see that it is sufficient to estimate the diameter of $\{r\} \times \omega_1$ (resp. $\{r\} \times \omega_2$) considered as a subset of $\mathbb{R}^+ \times U_\lambda$ (resp. $\mathbb{R}^+ \times U_{2\lambda}$) where this space is equipped with the warped product metric $r^{-2}dr^2 + r^{-2|\lambda|}du_\lambda^2(x_o)$ (resp. $r^{-2}dr^2+r^{-4|\lambda|}du_{2\lambda}^2(x_o)$) . If we apply Lemma 1.3 of [62] , then it follows that $\operatorname{diam}(B(z)) \leq C|\log(1/r(z))|$ as $r(z) \to 0$, $z \in Y$. Let $z \in Y$ and $t > 0$ be fixed. Then we claim that $\|e(z,.,t)\| \in L^1(Y) \cap L^2(Y)$. In view of (3.17), (3.21) and the estimation of $\operatorname{diam}(B(z'))$ just established, it is sufficient to prove that, for each $p \in \mathbb{N}$, the function $z' \to r(z')^p\exp(-d^2(z,z')/8t)$ is in $L^1(Y) \cap L^2(Y)$. But this follows from the fact that the volume of the set $\{z' \in Y \mid d(z,z') \leq R\}$, $R \in \mathbb{R}^+$, is bounded by Ce^{cR} for certain constants C,c [26, Lemma 4.1].

Thus, for each $\varphi \in L^2(Y,V)$, the integral

$$(H_t\varphi)(z) = \int_Y e(z,z',t)\cdot\varphi(z')dz'$$

is well defined. If we use formula (3.18) together with Fubini's theorem, then the right hand side can be rewritten as $R_\Gamma(h_t)$ where $R_\Gamma(h_t) = \int_G R_\Gamma(g) \otimes h_t(g)dg$. Let Q_Γ be the orthogonal projection of $L^2(\Gamma\backslash G) \otimes V$ onto its K-invariant subspace. According to the definition of Δ_μ , we have

$$\Delta_\mu = Q_\Gamma(-R_\Gamma(\Omega) \otimes \mathrm{Id} + \mathrm{Id} \otimes L)Q_\Gamma \quad .$$

Using computations similar to [15,p. 160] we obtain

$$e^{-t\Delta_\mu} = Q_\Gamma(e^{-tR_\Gamma(\Delta)} \otimes e^{t(2\mu(\Omega_K)-L)})\,Q_\Gamma \;, \qquad (3.23)$$

where Ω_K is the Casimir element of $\mathfrak{k}$ and Δ is defined by (3.12). On the other hand, by [70,Theorem 1.4] ,we have

$$R_\Gamma(p_t) = e^{-tR_\Gamma(\Delta)} \quad .$$

If we insert this on the right hand side of (3.23) and then apply (3.15), it follows that $R_\Gamma(h_t) = \exp(-t\Delta_\mu)$. Thus $e(z,z',t)$ is the kernel of the heat operator $\exp(-t\Delta_\mu)$. We summarize our results concerning the heat kernel on the cusp by

<u>PROPOSITION 3.24.</u> The kernel $e(z,z',t)$ of the heat operator $\exp(-t\Delta_\mu)$ is given by (3.18). e is contained in the space $C^\infty(Y\times Y\times \mathbb{R}^+, V\boxtimes V^*)$ and it satisfies

(i) $(\frac{\partial}{\partial t} + \Delta_\mu)e(z,z',t) = 0$ where Δ_μ acts either in the first or in the second variable.

(ii) For each $T,c > 0$ there exists a constant $C > 0$ such that

$$\| e(z,z',t)\| \leq C\, t^{-n/2}(r(z)r(z'))^{m/2} \exp(-\frac{d^2(z,z')}{8t})$$

for $0 < t \leq T$ and $z,z' \in Y_c$. Here $n = \dim Y$ and $m = m(\lambda)|\lambda|+m(2\lambda)|\lambda|$.

CHAPTER IV
THE NEUMANN LAPLACIAN ON THE CUSP

In this chapter, we shall investigate the selfadjoint operators which arise from certain locally invariant operators on cusps of rank one by imposing Neumann boundary conditions.

As in §3 let $\Delta_\mu : C^\infty(Y,\mathcal{V}) \to C^\infty(Y,\mathcal{V})$ denote the locally invariant operator defined by the element

$$-\Omega \otimes \mathrm{Id}_V + \mathrm{Id} \otimes L \in (\mathfrak{U}(\mathfrak{g}_{\mathbb{C}}) \otimes \mathrm{End}(V))^K . \tag{4.1}$$

Throughout this chapter we shall assume that L is such that Δ_μ is a positive operator. Since L will be fixed we shall suppress the letter L in the notation. Choose $b > 0$ and set $W = Y_b$ where $Y_b \subset Y$ is the submanifold defined by (2.13) . Furthermore, let $E = \mathcal{V}|Y_b$.

Let ∇ be the connection on $\mathcal{V}$ which is induced by the canonical invariant connection on $\tilde{\mathcal{V}}$. Let $\mathcal{D}'(Y,\mathcal{V})$ be the space of $\mathcal{V}$-valued distributions on Y . For $k \in \mathbb{N}$, we introduce the Sobolev space $H^k(Y,\mathcal{V})$ by

$$H^k(Y,\mathcal{V}) = \{\varphi \in \mathcal{D}'(Y,\mathcal{V}) \mid \nabla^l\varphi \in L^2(Y,\mathcal{V}) \text{ for all } l \leq k\} . \tag{4.2}$$

The norm of $\varphi \in H^k(Y,\mathcal{V})$ is defined by

$$\|\varphi\|_k = \sum_{l=0}^{k} \|\nabla^l\varphi\| .$$

Let

$$H^k(W,E) = \{\varphi \in W \mid \varphi \in H^k(Y,\mathcal{V})\} . \tag{4.3}$$

Let $\nabla^*\nabla$ be the connection Laplacian. It follows from (3.3) that

$$\Delta_\mu = \nabla^*\nabla + Q_\Gamma(\mathrm{Id} \otimes L_1)Q_\Gamma \tag{4.4}$$

where Q_Γ is the orthogonal projection of $L^2(\Gamma\backslash G) \otimes V$ onto the K-invariant subspace and L_1 belongs to $\mathrm{End}_K(V)$. On $H^1(W,E)$ we consider the quadratic form

$$q(\varphi) = \|\nabla\varphi\|^2 + (L\varphi,\varphi) , \tag{4.5}$$

where L is the algebraic operator determined by L_1. Let Δ_ν be the selfadjoint operator associated to q. To describe the domain of Δ_ν consider the space

$$\hat{C}_o^\infty(W,E) = \{\varphi \in C^\infty(W,E) \mid \operatorname{supp}\varphi \text{ compact and } \nabla_{\frac{\partial}{\partial r}}\varphi|\partial W = 0\} .$$

By restriction, Δ_μ induces a positive symmetric operator

$$\Delta_o : \hat{C}_o^\infty(W,E) \longrightarrow \hat{C}_o^\infty(W,E) .$$

It follows from (4.4) that for each $\varphi \in \hat{C}_o^\infty(W,E)$ one has

$$(\Delta_o\varphi,\varphi) = q(\varphi) .$$

Let $\mathbb{D}[\Delta_o]$ be the completion of $\hat{C}_o^\infty(W,E)$ with respect to the norm

$$|||\varphi|||^2 = (\Delta_o\varphi,\varphi) + ||\varphi||^2 .$$

Then we obtain $\mathbb{D}[\Delta_o] \subseteq H^1(W,E)$. On the other hand, it is easy to see that $\hat{C}_o^\infty(W,E)$ is a dense subsapce of $H^1(W,E)$. The argument is the same as one uses to prove that $\{f \in \overline{C}^\infty([a,b]) \mid f'(a) = f'(b) = 0\}$ is dense in $H^1([a,b])$. Thus $\mathbb{D}[\Delta_o] = H^1(W,E)$ and Δ_ν is Friedrichs' extension of Δ_o. Therefore, the domain $\mathbb{D}(\Delta_\nu)$ of Δ_ν is given by

$$\mathbb{D}(\Delta_\nu) = \mathbb{D}[\Delta_\nu] \cap \mathbb{D}(\Delta_o^*) \qquad (4.6)$$

and $\Delta_\nu = \Delta_o^* \restriction \mathbb{D}(\Delta_\nu)$ (c.f. Ch.I).

If $\varphi \in L^2(Y,V)$ then its constant term φ_o defined by (2.18) is again in $L^2(Y,V)$. Let $L_o^2(Y,V)$ be the space of cusp forms introduced by (2.19). We shall identify $L^2(W,E)$ with a subspace of $L^2(Y,E)$ extending sections by zero. Then we put

$$L_o^2(W,E) = L_o^2(Y,V) \cap L^2(W,E) . \qquad (4.7)$$

Now we recall the following facts from the theory of linear operators in a Hilbert space. Let T be a linear operator acting in a Hilbert space H with dense domain $\mathbb{D}(T)$. Assume that $H = H_1 \oplus H_2$ where H_1 and H_2 are orthogonal subspaces of H. Let P be the orthogonal projection of H onto H_1. The decomposition $H = H_1 \oplus H_2$ is called invariant under T if $P(\mathbb{D}(T)) \subset \mathbb{D}(T)$ and $T(\mathbb{D}(T) \cap H_i) \subset H_i$, $i=1,2$. In this case one can form the restriction T_i of T to H_i, $i=1,2$. The domain of T_i is given by $\mathbb{D}(T_i) = \mathbb{D}(T) \cap H_i$ and $T_iu = Tu$ if $u \in \mathbb{D}(T_i)$, $i=1,2$. If T is closed then T_1 and T_2 are closed operators [52,V §3.9,III,5.24]. If T is selfadjoint then

T_1 and T_2 are selfadjoint. With these observations we have

LEMMA 4.8. The decomposition

$$L^2(W,E) = L^2_o(W,E) \oplus (L^2_o(W,E))^{\perp}$$

into orthogonal subspace is invariant under Δ_ν. The restriction of Δ_ν to $L^2_o(W,E)$ and $(L^2_o(W,E))^{\perp}$, respectively, coincides with Friedrichs' extension of the restriction of Δ_o to $\hat{C}^\infty_o(W,E) \cap L^2_o(W,E)$ and $\hat{C}^\infty_o(W,E) \cap (L^2_o(W,E))^{\perp}$, respectively.

PROOF. Let $P : L^2(W,E) \to L^2_o(W,E)$ be the orthogonal projection. According to [52,V,§3.9] it is sufficient to prove that for each $\varphi \in \mathbb{D}(\Delta_\nu)$ one has $P\varphi \in \mathbb{D}(\Delta_\nu)$ and $\Delta_\nu P\varphi \in L^2_o(W,E)$. Let $f \in C^\infty(\Gamma\backslash G)$. Then it is obvious that $(R(\Omega)f)_o = R(\Omega)(f_o)$. Thus, if $\varphi \in \hat{C}^\infty_o(W,E)$ then we have $P\varphi \in \hat{C}^\infty_o(W,E)$ and $\Delta_o P\varphi \in L^2_o(W,E)$. Now the Lemma is an easy consequence of the description of the domain of Δ_ν by (4.6). Q.E.D.

Now we shall investigate the operator Δ_ν acting in the subspaces $L^2_o(W,E)$ and $(L^2_o(W,E))^{\perp}$, respectively.

PROPOSITION 4.9. Δ_ν has pure point spectrum in $L^2_o(W,E)$.

PROOF. We apply the method used by Lax and Phillips [57, p.206-207] in the two-dimensional case. Let $C^\infty_c(W,E)$ be the space of all C^∞ sections of E over W whose support is compact. Let $\varphi \in C^\infty_c(W,E) \cap L^2_o(W,E)$. We extend φ by zero to a section $\tilde{\varphi}$ of V over Y. Consider $\tilde{\varphi}$ as a function $\tilde{\varphi} : \Gamma\backslash G \to V$ which satisfies $\tilde{\varphi}(gk) = \mu(k)^{-1}\tilde{\varphi}(g)$, for all $k \in K$, $g \in G$. Let $v_1,\dots,v_q$ be an orthonormal basis for V and set $\varphi_j(g) = (\tilde{\varphi}(g),v_j)$. Then we have $\varphi_j \in L^2_o(\Gamma\backslash G)$, $j=1,\dots,q$. Let Δ_U be the Laplacian of $\Gamma\cap U\backslash U$ with respect to the left invariant metric on U. Let $Z_1,\dots,Z_l$ be an orthonormal basis for $\mathfrak{u}$ with respect to $(.,.)_\Theta$. Then we have

$$\Delta_U = - \sum_{j=1}^{l} R(Z_j)^2 .$$

Let $\{\psi_j\}_{j\in\mathbb{N}} \subset L^2(\Gamma\cap U\backslash U)$ be an orthonormal basis of eigenfunctions of Δ_U with eigenvalues $\{\mu_j\}_{j\in\mathbb{N}}$. We assume that $\mu_1 \le \mu_2 \le \dots$. For each i, $1 \le i \le q$, and $g \in G$, we may expand the function $u \mapsto \varphi_i(ug)$ on $\Gamma\cap U\backslash U$ in the basis $\{\psi_j\}_{j\in\mathbb{N}}$:

$$\varphi_i(ug) = \sum_{\mu_j>0} \varphi_{ij}(g)\psi_j(u)$$

$$\varphi_{ij}(g) = \int_{\Gamma\cap U\backslash U} \varphi_i(ug)\, \psi_j(u)\,du .$$

For $d > 0$ let

$$G_d = UA_dMK \ . \tag{4.10}$$

Let $C = (\mu_1)^{-1}$ and $d \geq b$. Then we get

$$\int_{G_d} |\varphi_i(g)|^2 dg \leq C \sum_{\mu_j > 0} \mu_j \int_{\Gamma_M \backslash M} \int_{A_d} \int_K |\varphi_{ij}(mak)|^2 e^{-2\rho(\log a)} dk\, da\, dm \leq \tag{4.11}$$

$$\leq C \int_{\Gamma \backslash G_d} \sum_{j=1}^{l} |R(-\mathrm{Ad}(g)Z_j)\varphi_i(g)|^2 dg \ .$$

Let $\omega \subset UM$ be a compact subset such that $\Gamma\omega = UM$ and set $S = \omega A_d K$. Let $X_1,\dots,X_p$ be an orthonormal basis for p with respect to $(.,.)_\theta$ and $Y_1,\dots,Y_q$ an orthonormal basis for k with respect to $(.,.)_\theta$. It is easy to see that there exist bounded functions α_{kj} and β_{hj} on S , $k=1,\dots,p$, $h=1,\dots,q$, such that

$$\mathrm{Ad}(g^{-1})Z_j = \Lambda(g)^{-1} \Big(\sum_{k=1}^{p} \alpha_{kj}(g)X_k + \sum_{h=1}^{q} \beta_{hj}(g)Y_h \Big) , \quad g \in S$$

$j=1,\dots,l$. $\Lambda(g)$ is the function defined by (2.6). By definition we have $\Lambda(g) \geq d$ on G_d . Finally, we observe that $R(Y_h)\tilde{\varphi}(g) = d\mu(-Y_h) \circ \tilde{\varphi}(g)$ and $\nabla_{X_k} = R(X_k)$. These observations combined with (4.11) imply that

$$\|\varphi\|^2 = \int_{\Gamma \backslash G_d} |\tilde{\varphi}(g)|^2 dg \leq \frac{C_1}{d^2} \int_{\Gamma \backslash G_d} (|\nabla\tilde{\varphi}(g)|^2 + |\tilde{\varphi}(g)|^2) dg \leq \frac{C_1}{d^2} \|\varphi\|_1^2 \tag{4.12}$$

The constant C_1 is independent of d and φ . Now observe that $C_c^\infty(W,E)$ is dense in $H^1(W,E)$. Therefore, (4.12) is valid for all $\varphi \in H^1(W,E) \cap L_o^2(W,E)$. For $d > b$ let $W_d = Y_b - Y_d$ and $E_d = E \restriction W_d$. By Rellich's theorem the embedding

$$i_d : H^1(W_d,E_d) \cap L_o^2(W_d,E_d) \longrightarrow L^2(W,E)$$

is compact. In view of (4.12), i_d converges strongly as $d \longrightarrow \infty$ to the embedding

$$i : H^1(W,E) \cap L_o^2(W,E) \longrightarrow L^2(W,E) \quad .$$

Hence i is compact. But $H^1(W,E) \cap L_o^2(W,E)$ is precisely the domain of the quadratic form defined by the restriction of Δ_ν to $L_o^2(W,E)$. This implies the proposition. Q.E.D.

Now we shall investigate the restriction of Δ_ν to the subspace $(L^2_o(W,E))^\perp$. First we observe that the orthogonal complement of $L^2_o(Y,V)$ in $L^2(Y,V)$ is the space of sections $\varphi \in L^2(Y,V)$ which satisfy $\varphi = \varphi_o$. Thus, according to §2, we get

$$(L^2_o(Y,V))^\perp = (L^2((\Gamma\cdot U)\backslash G) \otimes V)^K . \tag{4.13}$$

Next we determine the action of Δ_μ on $(C^\infty((\Gamma\cdot U)\backslash G) \otimes V)^K$. Set $M_1 = MA$ and $dm_1 = e^{-2\rho(\log a)}dmda$. Let f be a locally bounded measurable function $\Gamma\backslash G \rightarrow V$ which satisfies $f(gk) = \sigma(k)^{-1}f(g)$, $k \in K$, and put

$$(\pi f)(m_1) = f_o(m_1) , \quad m_1 \in M_1 . \tag{4.14}$$

Since $\Gamma_M = (\Gamma\cdot U) \cap M$, it follows that $f_o(\gamma m_1) = f_o(m_1)$ for $\gamma \in \Gamma_{M_1}$ and $m_1 \in M_1$. Therefore, πf is a locally bounded measurable function on $\Gamma_M\backslash M_1$ with values in V satisfying $\pi f(m_1k) = \sigma(k)^{-1}\pi f(m_1)$ for $k \in K_M$ and π induces an isomorphism

$$\pi : (C^\infty((\Gamma\cdot U)\backslash G) \otimes V)^K \xrightarrow{\sim} (C^\infty(\Gamma_M\backslash M_1) \otimes V)^{K_M} . \tag{4.15}$$

Choose a θ-stable Cartan subalgebra $\mathfrak{j}$ of $\mathfrak{g}$ containing the Lie algebra $\mathfrak{a}$ of A. By Φ we denote the set of nonzero roots of the pair $(\mathfrak{g}_{\mathbb{C}}, \mathfrak{j}_{\mathbb{C}})$. For $\alpha \in \Phi$ choose $H_\alpha \in \mathfrak{j}_{\mathbb{C}}$ and $X_\alpha \in \mathfrak{g}^\alpha_{\mathbb{C}}$ satisfying the usual conditions, i.e. $B(X_\alpha, X_{-\alpha}) = 1$, $B(H,H_\alpha) = \alpha(H)$ for all $H \in \mathfrak{j}_{\mathbb{C}}$ and $[X_\alpha, X_{-\alpha}] = H_\alpha$. Let $H_1,\dots,H_k,\dots,H_l$ be a basis for $\mathfrak{j}$ such that $H_1 \in \mathfrak{a}$, $H_1,\dots,H_k$ is a basis for $\mathfrak{j}_{\mathfrak{p}} = \mathfrak{j} \cap \mathfrak{p}$ and $H_{k+1},\dots,H_l$ is a basis for $\mathfrak{j}_{\mathfrak{k}} = \mathfrak{j} \cap \mathfrak{k}$ over $\mathbb{R}$. Suppose further that $B(H_i,H_j) = \delta_{ij}$ for $1 \le i,j \le k$ and $B(H_i,H_j) = -1$ for $k+1 \le i,j \le l$. Then

$$\Omega = \sum_{i=1}^{k} H_i^2 - \sum_{j=k+1}^{l} H_j^2 + \sum_{\alpha>0} (X_\alpha X_{-\alpha} + X_{-\alpha}X_\alpha) . \tag{4.16}$$

Set $H = H_1$ and let $\Phi_H = \{\alpha \in \Phi \mid \alpha(H) = 0\}$. If $\alpha \in \Phi_H$ then $X_\alpha, X_{-\alpha} \in z(\mathfrak{a}_{\mathbb{C}})$, the centralizer of $\mathfrak{a}_{\mathbb{C}}$ in $\mathfrak{g}_{\mathbb{C}}$. On the other hand, if $\alpha(H) \neq 0$ and $\alpha > 0$ then we have $\mathfrak{g}^\alpha_{\mathbb{C}} \subset \mathfrak{u}_{\mathbb{C}}$ and $\mathfrak{u}_{\mathbb{C}}$ is spanned by all these root spaces. By $\mathfrak{Z}(\mathfrak{m}_{\mathbb{C}})$ we denote the center of the universal enveloping algebra of $\mathfrak{m}_{\mathbb{C}}$. Note that $\mathfrak{j}$ is also a Cartan subalgebra of $\mathfrak{m}_1 = \mathfrak{m} \oplus \mathfrak{a}$. Finally, observe that $X_{-\alpha}X_\alpha = X_\alpha X_{-\alpha} - H_\alpha$. Using these observations together with (4.16) it is easy to see that there exists a unique selfadjoint element $\Omega_M \in \mathfrak{Z}(\mathfrak{m}_{\mathbb{C}})$ such that

$$\Omega = H^2 - 2\rho(H)H + \Omega_M \pmod{\mathfrak{u}\,\mathfrak{U}(\mathfrak{g}_{\mathbb{C}})} . \tag{4.17}$$

It follows from the definition of Ω_M that

$$\Omega_M = \sum_{i=2}^{k} H_i^2 - \sum_{j=k+1}^{l} H_j^2 + \sum_{\substack{\alpha>0\\ \alpha(H)=0}} (X_\alpha X_{-\alpha} + X_{-\alpha} X_\alpha) \quad .$$

Let $\mathfrak{m} = \mathfrak{k}_M \oplus \mathfrak{p}_M$ be the Cartan decomposition of $\mathfrak{m}$ with respect to θ_M. Choose a basis $X_1,\ldots,X_t$ for $\mathfrak{m}$ such that $X_1,\ldots,X_q$ is an orthonormal basis for $\mathfrak{k}_M$ with respect to $-B|\mathfrak{k}_M \times \mathfrak{k}_M$ and $X_{q+1},\ldots,X_t$ is an orthonormal basis for $\mathfrak{p}_M$ with respect to $B|\mathfrak{p}_M \times \mathfrak{p}_M$. Then, with respect to this basis of $\mathfrak{m}$, Ω_M is given by

$$\Omega_M = -\sum_{j=1}^{q} X_j^2 + \sum_{j=q+1}^{t} X_j^2 \quad . \tag{4.18}$$

Set

$$\gamma(\Omega) = H^2 - 2\rho(H)H + \Omega_M \quad . \tag{4.19}$$

Then $\gamma(\Omega) \in \mathfrak{Z}(\mathfrak{g}_{\mathbb{C}}) \otimes S(\mathfrak{a}_{\mathbb{C}})$ and we observe that $\gamma(\Omega)$ is closely related to the image of Ω under the canonical injective homomorphism $\mathfrak{Z}(\mathfrak{g}_{\mathbb{C}}) \to \mathfrak{Z}(\mathfrak{m}_{\mathbb{C}}) \otimes S(\mathfrak{a}_{\mathbb{C}})$ (see [44,I,§6] for its definition).

Let $f \in C^\infty(\Gamma\backslash G)$ and assume that $f(ug) = f(g)$ for all $u \in U$. Then (4.17) implies that $(R(\Omega)f)(m_1) = (R(\gamma(\Omega))f)(m_1)$, $m_1 \in M_1$. Therefore we get

$$\pi((R_\Gamma(\Omega) \otimes \mathrm{Id})\phi) = (R'_{\Gamma_M}(\gamma(\Omega)) \otimes \mathrm{id})\pi(\phi) \ , \tag{4.20}$$

$\phi \in (C^\infty(\Gamma\backslash G) \otimes V)^K$ and R'_{Γ_M} denotes the right regular representation of M_1 on $C^\infty(\Gamma_M\backslash M_1)$.

As in §2, we shall identify A with $\mathbb{R}^+$ via the map $a \in A \to \exp(\lambda(\log a)/|\lambda|)$. Let $\mu_M = \mu\restriction K_M$ and consider the associated locally homogeneous vector bundle V_M over $\Gamma_M\backslash X_M$ where $X_M = M/K_M$. We lift V_M to a vector bundle over $\mathbb{R}^+ \times \Gamma_M\backslash X_M$. Then we have

$$(C^\infty(\Gamma_M\backslash M_1) \otimes V)^{K_M} \cong C^\infty(\mathbb{R}^+ \times \Gamma_M\backslash X_M, V_M) \ . \tag{4.21}$$

Moreover, the operator $R(H)$ acting on $C^\infty(A)$ corresponds to $r\frac{d}{dr}$, and, if $H_0 \in \mathfrak{a}$ is such that $\lambda(H_0) > 0$ and $\|H_0\| = 1$, then one has $|\lambda| = \lambda(H_0)$ and $2\rho(H_0) = m$. Since MA is the centralizer of A in G we can conclude that

$$R'_{\Gamma_M}(\gamma(\Omega)) = r^2\frac{\partial^2}{\partial r^2} + (1-m)r\frac{\partial}{\partial r} + R_{\Gamma_M}(\Omega_M) \ , \tag{4.22}$$

R_{Γ_M} being the right regular representation of M on $C^\infty(\Gamma_M\backslash M)$. Set

$$\Delta_M = -R_{\Gamma_M}(\Omega_M) \otimes \mathrm{Id} + \mathrm{Id} \otimes L \tag{4.23}$$

and let

$$T = -r^2 \frac{\partial^2}{\partial r^2} + (1-m) r \frac{\partial}{\partial r} + \Delta_M \tag{4.24}$$

acting on $C^\infty(\mathbb{R}^+ \times \Gamma_M \backslash X_M, V_M)$. Moreover, let

$$\tilde{\pi} : (C^\infty((\Gamma \cdot U)\backslash G)) \otimes V)^K \xrightarrow{\sim} C^\infty(\mathbb{R}^+ \times \Gamma_M \backslash X_M, V_M)$$

be the composition of (4.15) and (4.21). Using (4.20), we obtain

$$\tilde{\pi} \circ \Delta_\mu = T \circ \tilde{\pi} . \tag{4.25}$$

Now we can continue with the investigation of Δ_ν acting on $(L_o^2(W,E))^\perp$. Using the map (4.14) and the identification of A with $\mathbb{R}^+$, we obtain an isomorphism

$$(L_o^2(W,E))^\perp \xrightarrow{\sim} L^2([b_1,\infty) \times \Gamma_M \backslash X_M, V_M) \tag{4.26}$$

where $b_1 = b^{1/|\lambda|}$. Since $\Omega_M \in \mathfrak{Z}(m_\mathbb{C})$ is a selfadjoint element it follows from (4.23) that Δ_M is an elliptic differential operator whose closure in $L^2(\Gamma_M \backslash X_M, V_M)$ is selfadjoint. Let ∇_M be the connection on V_M which is obtained by pushing down the canonical invariant connection on $\tilde{V}_M$ and let $\nabla_M^* \nabla_M$ be the connection Laplacian on V_M associated to the Hermitian structure of V_M and the left invariant metric on X_M. If we employ the argument which we used to prove formula (3.3), then it follows from (4.18) that

$$\Delta_M = \nabla_M^* \nabla_M + L \tag{4.27}$$

where L is the algebraic operator defined by L. Hence, Δ_M is bounded from below. Now consider the operator

$$T_o = -r^2 \frac{d^2}{dr^2} + (m-1) r \frac{d}{dr} \tag{4.28}$$

with $\mathbb{D}(T_o) = \hat{C}_c^\infty([b_1,\infty)) = \{f \in C_c^\infty([b_1,\infty)) \mid f'(b_1) = 0\}$. Let $(T_o)_F$ be Friedrichs' extension of T_o. Then it follows from (4.25) that the restriction of Δ_ν to $(L_o^2(W,E))^\perp$ corresponds under the isomorphism (4.26) to the operator

$$T_\nu = (T_o)_F + \Delta_M . \tag{4.29}$$

Is it easy to determine the spectral resolution of this operator. The constants are the only eigenfunctions of $(T_o)_F$ and the restriction of $(T_o)_F$ to the orthogonal complement to $\mathbb{C}$ is absolutely continuous. Generalized eigenfunctions are

$$\eta(r,s) = r^s + \frac{s}{s-m} b_1^{2s-m} r^{m-s} \quad , \ s \in \mathbb{C} \ . \tag{4.30}$$

Acting on a compact manifold, Δ_M has pure point spectrum. The spectrum of Δ_M is contained in a half line $[c,\infty)$. Let $\{\phi_k\}_{k\in\mathbb{N}} \subset L^2(\Gamma_M\backslash X_M, V_M)$ be an orthonormal basis of eigenfunctions of Δ_M with eigenvalues $\mu_1 \le \mu_2 \le \dots \longrightarrow \infty$. We shall identify $L^2(\Gamma_M\backslash X_M, V_M)$ with the subspace $\mathbb{C} \otimes L^2(\Gamma_M\backslash X_M, V_M)$ of $L^2([b_1,\infty) \times \Gamma_M\backslash X_M, V_M)$. T_ν has pure point spectrum in this subspace. Let $L^2_c([b_1,\infty) \times \Gamma_M\backslash X_M, V_M)$ be the orthogonal complement to $L^2(\Gamma_M\backslash X_M, V_M)$ in $L^2([b_1,\infty) \times \Gamma_M\backslash X_M, V_M)$. Put

$$\eta_k(r,x,s) = \eta(r,s)\ \phi_k(x) \quad , \quad k \in \mathbb{N} \quad , \ s \in \mathbb{C} \ . \tag{4.31}$$

These sections can be considered as generalized eigenfunctions of T_ν . For $\varphi \in C_c^\infty([b_1,\infty) \times \Gamma_M\backslash X_M, V_M)$ and $k \in \mathbb{N}$ we define $\hat{\varphi}(k) \in L^2(\mathbb{R}^+)$ by

$$\hat{\varphi}(\zeta,k) = \int_{b_1}^{\infty} \int_{\Gamma_M\backslash X_M} (\varphi(r,x),\ \eta_k(r,x,\tfrac{m}{2} + i\zeta))\, dx \frac{dr}{r^{m+1}} \ . \tag{4.32}$$

The mapping $\varphi \longmapsto \hat{\varphi}$ can be extended to a bounded linear operator.

$$J : L^2([b_1,\infty) \times \Gamma_M\backslash X_M, V_M) \longrightarrow L^2(\mathbb{R}^+;\ell^2) \ . \tag{4.33}$$

J^*J is the orthogonal projection onto $L^2_c([b_1,\infty) \times \Gamma_M\backslash X_M, V_M)$ and J induces an isometry

$$L^2_c([b_1,\infty) \times \Gamma_M\backslash X_M, V_M) \xrightarrow{\ \sim\ } L^2(\mathbb{R}^+;\ell^2,\tfrac{d\zeta}{2\pi})$$

Moreover, we have

$$JT_\nu J^* = \tilde{B} \quad , \tag{4.34}$$

where $\tilde{B}$ is the operator defined by $(\tilde{B}\hat{\varphi}(\zeta,k) = (\frac{m^2}{4} + \zeta^2 + \mu_k)\hat{\varphi}(\zeta,k)$. If $\hat{\varphi} \in C_o^\infty(\mathbb{R}^+;\ell^2)$ then $J^*\hat{\varphi}$ is given by

$$(J^*\hat{\varphi})(r,x) \sum_{k=1}^{\infty} \frac{1}{2\pi}\Big(\int_o^\infty \eta(r,\tfrac{m}{2} - i\zeta)\hat{\varphi}(\zeta,k)\,d\zeta\Big)\,\phi_k(x) \ . \tag{4.35}$$

Let

$$V : \bigoplus_{k=1}^{\infty} L^2([\tfrac{m^2}{4} + \mu_k,\infty)) \longrightarrow L^2(\mathbb{R}^+;\ell^2)$$

be defined by

$$(Vf)(\zeta,k) = \sqrt{2\lambda}\ f_k(\tfrac{m^4}{4} + \mu_k + \zeta^2) \ . \tag{4.36}$$

V is an isometry and we have $V^*\tilde{B}V = B$ where B is the operator which is given by

$$(Bf)_k(\zeta) = \zeta f_k(\zeta) \quad , \quad f \in \bigoplus_{k=1}^{\infty} L^2([\tfrac{m^2}{4}+\mu_k,\infty)) \quad . \tag{4.37}$$

Thus V^*J gives the spectral resolution of T_ν acting on $L^2_c([b_1,\infty)\times \Gamma_M\backslash X_M, V_M)$. We summarize our results concerning the spectral resolution of Δ_ν by

THEOREM 4.38. (i) The spectrum of Δ_ν consists of a point spectrum σ_p and an absolutely continuous spectrum σ_{ac} . The point spectrum consists of eigenvalues of finite multiplicity with ∞ as the only point of accumulation. Let μ_1 be the lowest eigenvalue of the operator Δ_M defined by (4.25) and let $m = m(\lambda)|\lambda| + 2m(2\lambda)|\lambda|$. Then we have $\sigma_{ac} = [m^2/4+\mu_1,\infty)$.

(ii) The subspace of $L^2(W,E)$ which is spanned by the eigenfunctions of Δ_ν is equal to $L^2_o(W,E) \oplus L^2(\Gamma_M\backslash X_M, V_M)$. The absolutely continuous part of Δ_ν is unitarily equivalent to the multiplication operator on $\bigoplus_{k=1}^{\infty} L^2([m^2/4+\mu_k,\infty),\frac{d\zeta}{2\pi})$.

The number of cuspidal eigenvalues of Δ_ν which are less than λ can be estimated as in [27] . It is bounded by a constant multiple of $\lambda^{n/2}$ where $n=\dim X$. The remaining part of the discrete spectrum are the eigenvalues of Δ_M . It follows from (4.27) that the principal symbol of Δ_M is given by $\sigma(\Delta_M)(x,\xi) = \langle\xi,\xi\rangle_x I$, where $\langle .,.\rangle_x$ is the Riemannian metric in $T^*_x(\Gamma_M\backslash X_M)$. Using the parametrix method one can construct the fundamental solution $K(x,x',t)$ of the heat equation for Δ_M (c.f. [76,§2]) . The kernel satisfies

$$K(x,x,t) = (4\pi t)^{-q/2} I_x + O(t^{-(q-1)/2})$$

as $t \to 0$, $t > 0$, $q = \dim X_M$ (c.f. [76,Lemma 2.4]) . Using a Tauberian theorem [79,p.192] we obtain

PROPOSITION 4.39. Let $N_\nu(\lambda)$ be the number of eigenvalues, counted to multiplicity, of Δ_ν which are less than λ . There exists a constant $C > 0$ such that

$$N_\nu(\lambda) \leq C\lambda^{n/2} \quad , \quad \lambda > 0 \quad .$$

CHAPTER V
MANIFOLDS WITH CUSPS OF RANK ONE

In this chapter we shall consider Riemannian manifolds which in the complement of a compact set are isometric to the disjoint union of a finite number of cusps of rank one. On these manifolds we shall investigate elliptic differential operators which are locally invariant on the cusps. Of particular interest are those operators which on each cusp coincide with one of the locally invariant operators studied in the previous sections. Our first purpose is to get information about the spectral resolution of these operators. The absolutely continuous part of the spectrum of these operators will be identified, up to unitary equivalence.

To begin with we make the following

DEFINITION 5.1. An oriented Riemannian manifold X is called a manifold with cusps of rank one if X is complete and if it has a decomposition $X = X_o \cup X_1 \cup \ldots \cup X_h$ satisfying the following conditions

(i) X_o is a compact manifold with boundary. If $i,j \geq 1$ and $i \neq j$, then $X_i \cap X_j = \emptyset$. Moreover $X_o \cap X_i = \partial X_i$ for $i \geq 1$.

(ii) For each i, $1 \leq i \leq h$, there exists a cusp of rank one, say Y_i, and $b_i > 0$ such that X_i is isometric to the manifold Y_{i,b_i} defined by (2.13).

Before going into details we shall discuss several examples.

EXAMPLE 1. Let G and K be as above. A discrete subgroup Γ of G is called a lattice of rank one if it has finite covolume and each Γ-percuspidal split parabolic subgroup of G is of rank one (see [66,Ch.2] for the definitions). Assume that Γ is torsion free. Let (P,S) be any Γ-percuspidal split parabolic subgroup of G with Θ-stable split component A and Langlands decomposition $P = UAM$. According to the definition of P, one has $\Gamma \cap P \subset S$ and $\Gamma \cap S\backslash S$ is compact [66,p. 35]. Thus $\Gamma \cap P\backslash G/K$ is a cusp of rank one in the sense of Definition 2.2. For $b > 0$ set $Y_b(P) = \Gamma \cap P\backslash U \cdot A_b \cdot M/K_M$. It follows from [67, Theorem 2.1] that there exists $b > o$ and Γ-percuspidal split parabolic subgroups $P_1,\ldots,P_r$ of G such that, for each i,

$i=1,\dots,r$, the natural mapping $Y_b(P_i) \to \Gamma\backslash G/K$ is an embedding, the submanifolds $Y_b(P_1),\dots,Y_b(P_r)$ are mutually disjoint and $\Gamma\backslash G/K - \bigcup_{i=1}^{r} Y_b(P_i)$ is compact. This means precisely that $\Gamma\backslash G/K$ is a manifold with cusps of rank one in the sense of Definition 5.1. These are locally symmetric spaces with "point-like" cusps as discussed by Selberg in his talk on the ICM in Stockholm. Note that, in particular, each locally symmetric space of finite volume and strictly negative sectional curvatures is a manifold with cusps of rank one. Other examples of rank one lattices can be obtained as follows. Let F be an algebraic number field of degree $n = [F:\mathbb{Q}]$ and denote by $\mathcal{O}_F$ its ring of integers. Let r_1 be the number of real places and $2r_2$ the number of complex places of F . The group $SL(2,\mathcal{O}_F)$ can be considered as a discrete subgroup of the Lie group

$$G = SL(2,\mathbb{R})^{r_1} \times SL(2,\mathbb{C})^{r_2}$$

by sending $\begin{pmatrix} a & b \\ c & d \end{pmatrix} \in SL(2,\mathcal{O}_F)$ to .

$$\left(\begin{pmatrix} a^{(1)} & b^{(1)} \\ c^{(1)} & d^{(1)} \end{pmatrix},\dots,\begin{pmatrix} a^{(n)} & b^{(n)} \\ c^{(n)} & d^{(n)} \end{pmatrix}\right) \in G ,$$

where $x \to x^{(j)}$ denotes the j-th embedding of F in $\mathbb{C}$. Let Γ be a subgroup of $SL(2,\mathcal{O}_F)$ of finite index. Then Γ is a rank one lattice in G (see [38]). We choose the maximal compact subgroup as $K = SO(2)^{r_1} \times SU(2)^{r_2}$. The corresponding symmetric space is $G/K = H^{r_1} \times (H^3)^{r_2}$, where H is the upper half-plane and H^3 the hyperbolic three-space. If $r_2=0$ then we get a Hilbert modular variety $\Gamma\backslash H^n$.

<u>EXAMPLE 2</u>. We return to Example 2.3. Let F be a totally real number field of degree n over $\mathbb{Q}$, $\mathbf{M}$ a complete $\mathbb{Z}$-module of F and $\mathbf{V} \subset U_{\mathbf{M}}^{+}$ a subgroup of finite index. Let Γ be a group of matrices $\begin{pmatrix} \varepsilon^{1/2} & \varepsilon^{-1/2}\mu \\ 0 & \varepsilon^{-1/2} \end{pmatrix}$ such that the sequence

$$0 \longrightarrow \mathbf{M} \longrightarrow \Gamma \longrightarrow \mathbf{V} \longrightarrow 1$$

is exact. The group Γ acts freely and properly discontinuously on the product of upper half-planes H^n and $Y = \Gamma\backslash H^n$ is a cusp of rank one. For $d > 0$ set

$$W(d) = \{z \in H^n \mid \mathrm{Im}(z_1)\dots\mathrm{Im}(z_n) \geq d\} .$$

$W(d)$ is invariant under Γ and $\Gamma\backslash W(d)$ is the submanifold Y_d of Y defined by (2.13), which we continue to call cusp. The manifold $\Gamma\backslash W(d)$ inherits the standard parallelization of H^n given by the coordinates $x_1,y_1,\ldots,x_n,y_n$ (with $z_k = x_k+iy_k$). This parallelization is respected by Γ if we use unit vectors with respect to the invariant metric of H^n. Thus the stable tangent bundle of $N = \partial(\Gamma\backslash W(d))$ has also a canonical parallelization. In particular, there exists a compact oriented C^∞-manifold X_o with $\partial X_o = N$. Let X be the C^∞-manifold obtained by gluing X_o to Y_d along their common boundary. We choose on X a Riemannian metric which coincides with the given one on Y_d. Then X is a manifold with a cusp of rank one. This manifold will be important for our proof of Hirzebruch's conjecture.

EXAMPLE 3. Of course, the previous example can be generalized. Let $Y = \Gamma\backslash G/K$ be a cusp of rank one and $N = \Gamma\backslash Z$ where $Z = S/S\cap K$. As soon as we know that N bounds a compact oriented C^∞-manifold we can repeat the construction of example 2 and obtain a manifold with a cusp of rank one. If we assume that the dimension of Y is even, then at least a multiple of N is a boundary. A large class of manifolds with a single cusp of rank one can be obtained as follows. Let $D = G/K$ be a n-dimensional bounded symmetric domain and assume that $G = \mathrm{Aut}(D)^o$ is an algebraic group defined over $\mathbb{Q}$ with $\mathbb{Q}$-rank $G = 1$. Let Γ be a neat arithmetic subgroup of G. Let F be a rational boundary component of D and let $P(F) = \{g\in G \mid gF = F\}$. Then $P(F)$ is a parabolic subgroup of G defined over $\mathbb{Q}$. The space $(\Gamma\cap P(F)\backslash D)\cup(\Gamma\cap P(F)\backslash F)$ has a natural analytic structure [13]. For $c>0$ let D_c be the subspace of D which is defined by (2.10) with respect to $P(F)$. Let X_o be the manifold obtained by resolving the singularities of $(\Gamma\cap P(F)\backslash D_c)\cup(\Gamma\cap P(F)\backslash F)$. Gluing X_o to the cusp $\Gamma\cap P(F)\backslash D_c$ along the common boundary we obtain a manifold X with a cusp of rank one.

For simplicity we shall assume throughout this paper that the manifolds we are considering have a single cusp. Such a manifold X has a dacomposition $X = X_o \cup Y_b$, $b>0$, where X_o is a compact manifold with boundary and Y_b is a cusp of rank one. After rescaling the metric we may assume that $b=1$. Clearly, vector bundles and differential operators on manifolds with cusps have to respect the locally symmetric structure at infinity.

DEFINITION 5.2. Let $X = X_o \cup Y_1$ be a manifold with a cusp of rank one. A vector bundle E over X is called locally homogeneous at infinity if there exists a locally homogeneous vector bundle $\mathcal{E}$ over $Y = \Gamma\backslash G/K$ such that the restrictions $E|Y_1$ and $\mathcal{E}|Y_1$ are isomorphic through a bundle map which covers the identity of Y_1. If E is equipped with a Hermitian metric, then we assume that $\mathcal{E}$ carries a Hermitian metric induced by an invariant metric on the homogeneous vector bundle $\tilde{\mathcal{E}}$ over $\tilde{Y}$ and the bundle map which identifies $E|Y_1$ and $\mathcal{E}|Y_1$ has to respect the metrics.

If E is any vector bundle over X we shall denote by $C^\infty(X,E)$ (resp. $C_c^\infty(X,E)$) the space of all C^∞-sections (resp. C^∞-sections with compact supports) of E. If E has a Hermitian metric, then we shall denote by $L^2(X,E)$ the Hilbert space of all L^2-sections of E. The inner product is defined via the Hermitian metric of E and the Riemannian measure on X.

Let E,F be vector bundles over X which are locally homogeneous at infinity and let $\mathcal{E},\mathcal{F}$ be the corresponding locally homogeneous vector bundles over the cusp Y associated to X. A differential operator

$$D: C^\infty(X,E) \longrightarrow C^\infty(X,F)$$

will be called locally invariant at infinity if there exists a locally invariant differential operator

$$\mathcal{D}: C^\infty(Y,\mathcal{E}) \longrightarrow C^\infty(Y,\mathcal{F})$$

such that $D = \mathcal{D}$ on Y_1. From now on we shall assume that all vector bundles are equipped with a Hermitian metric.

Now we shall introduce an important class of first-order differential operators on manifolds with cusps of rank one.

DEFINITION 5.3. Let X be a manifold with a cusp of rank one. A generalized Dirac operator on X is a first-order elliptic differential operator $D: C^\infty(X,E) \longrightarrow C^\infty(X,E)$ which satisfies the following properties:

(i) D is formally selfadjoint and locally invariant at infinity.

(ii) Let $\mathcal{D}: C^\infty(Y,\mathcal{E}) \longrightarrow C^\infty(Y,\mathcal{E})$ be the locally invariant differential operator associated to D and assume that V is the unitary K-module associated to $\tilde{\mathcal{E}}$. Then there exists $L \in \mathrm{End}_K(V)$ such that

$$\mathcal{D}^2 = -R(\Omega) \otimes \mathrm{Id}_V + \mathrm{Id} \otimes L . \tag{5.4}$$

This definition is justified for the following reason. Let G and K be as above and assume that the isotropy representation $Ad : K \to SO(p)$ of the tangent bundle of G/K lifts to a homomorphism $\alpha : K \to Spin(p)$. Then G/K has a G-invariant spin structure. Let s be the basic spin representation of $Spin(p)$ on S and denote by $\tilde{S}$ the homogeneous vector bundle associated to the representation $s \circ \alpha$ of K on S. Let $\tilde{V}$ be another homogeneous vector bundle over G/K defined by a finite-dimensional unitary representation of K on V. Let $X_1, \ldots, X_p$ be an orthonormal basis for p and set

$$Z = \sum_{i=1}^{p} X_i \otimes Id_V \otimes c(X_i) \in \mathcal{U}(g_{\mathbb{C}}) \otimes End(V \otimes S) \quad ,$$

where $c(X_i)$ denotes the Clifford multiplication by $X_i \in p$. Z is invariant under K and therefore, it defines a G-invariant first-order differential operator

$$\tilde{D}_V : C^\infty(G/K, \tilde{V} \otimes \tilde{S}) \longrightarrow C^\infty(G/K, \tilde{V} \otimes \tilde{S}) \quad ,$$

called Dirac operator with coefficients in $\tilde{V}$. $\tilde{D}_V$ is elliptic and formally selfadjoint. Using the results of Parthasarathy [69,§3] , it follows that the spinor Laplacian $\tilde{D}_V^2$ satisfies condition (5.4) . Let Γ be a lattice of rank one in G and assume that Γ is torsion free. Then the locally invariant operator

$$D_V : C^\infty(\Gamma \backslash G/K, V \otimes S) \longrightarrow C \; (\Gamma \backslash G/K, V \otimes S)$$

satisfies all conditions of Definition 5.3.

We observe that geometrically interesting operators fall into the class of operators singled out by Definition 5.3. First consider the Laplacian Δ acting on the space $\Lambda^*(X)$ of differential forms. Let $Ad_p : K \to GL(p_{\mathbb{C}})$ be the restriction of the adjoint representation. The homogeneous vector bundle $\Lambda^* T^*(G/K)$ is associated to the representation $\Lambda^* Ad_p^*$ of K on $\Lambda^* p_{\mathbb{C}}^*$. According to Kuga's Lemma [58,p.385], the Laplacian on $\Lambda^*(G/K)$ coincides with the restriction of $-R(\Omega) \otimes Id$ to the K-invariant subspace of $C^\infty(G) \otimes \Lambda^* p_{\mathbb{C}}^*$. Thus

$$d + \delta : \Lambda^*(X) \to \Lambda^*(X)$$

is a generalized Dirac operator. Another important example is the Dirac operator itself [12,§5] . To define the Dirac operator one has to assume that X is a spin manifold, i.e. the structure group of the principal tangent bundle of X can be lifted from $SO(n)$ to $Spin(n)$. We choose a definite spin structure. Then let S be the vector bundle over X associated to the basic spin representation.

The restriction of S to Y_1 is given by the restriction of the corresponding locally homogeneous vector bundle $\mathcal{S}$ over Y. The Dirac operator D is a first-order elliptic differential operator

$$D : C^\infty(X,S) \longrightarrow C^\infty(X,S) \quad .$$

Since the Dirac operator is canonically associated to the metric and the spin structure, it follows that D is locally invariant at infinity. Finally, we have seen above that the Dirac operator of the symmetric space G/K satisfies the condition (5.4). Thus D is a generalized Dirac operator in the sense of Definition 5.3.

Now assume that $\dim X$ is even. Then the basic spin representation breaks up in two half-spin representations $s^\pm$ of $\mathrm{Spin}(n)$. Let $S^\pm$ be the associated vector bundles over X. Then D maps $C^\infty(S^\pm)$ to $C^\infty(S^\mp)$ and, by restriction, we obtain a pair of formally adjoint operators

$$D^\pm : C^\infty(X,S^\pm) \longrightarrow C^\infty(X,S^\mp) \quad .$$

The operators $D^\pm$ are called chiral Dirac operators. With respect to the splitting $C^\infty(S) = C^\infty(S^+) \oplus C^\infty(S^-)$, the Dirac operator D has the form

$$D = \begin{pmatrix} 0 & D^- \\ D^+ & 0 \end{pmatrix} \quad .$$

We generalize this by

<u>DEFINITION 5.3'</u>. A differential operator

$$Q : C^\infty(X,E) \longrightarrow C^\infty(X,F)$$

is called a generalized chiral Dirac operator, if the operator

$$D = \begin{pmatrix} 0 & Q^* \\ Q & 0 \end{pmatrix}$$

acting on $C^\infty(X,E) \oplus C^\infty(X,F)$ is a generalized Dirac operator in the sense of Definition 5.3.

Now we shall study the basic properties of generalized Dirac operators. Let

$$D : C^\infty(X,E) \longrightarrow C^\infty(X,E)$$

be a generalized Dirac operator and let

$$\mathcal{D} : C^\infty(Y,\mathcal{E}) \longrightarrow C^\infty(Y,\mathcal{E})$$

be the associated locally invariant operator. Assume that $\mathcal{E}$ is given by the isotropy representation σ of K on V. Since $\mathcal{D}$ is a first-order operator, it has a simple description:

LEMMA 5.5. Let $X_1,\dots,X_n$ be an orthonormal basis for p. Then there exist $A_j \in \mathrm{End}(V)$, $j=1,\dots,n$, and $A_o \in \mathrm{End}_K(V)$ such that

$$\tilde{\mathcal{D}} = \sum_{j=1}^{n} R(X_j) \otimes X_j + A_o .$$

The principal symbol $\sigma_{\tilde{\mathcal{D}}}$ of $\tilde{\mathcal{D}}$ is an element of $\mathrm{Hom}_K(p^*,\mathrm{End}(V))$ and it is given by

$$\sigma_{\tilde{\mathcal{D}}}(df) = \sum_{j=1}^{n} (X_j f)A_j \quad , \; f \in C^\infty(Y) .$$

For the proof see [60,p.6] .

LEMMA 5.6. Let $p : T^*X \to X$ be the projection and $\sigma_D : p^*E \to p^*E$ the symbol of D. For $f \in C^\infty(X)$ and $\varphi \in C^\infty(X,E)$ one has

$$D(f\varphi)(x) = \sigma_D(df_x,x)(\varphi(x)) + f(x)D(\varphi)(x) .$$

PROOF. Let $U \subset X$ be a coordinate chart such that $E|U \simeq U \times V$. Then we have

$$D = \sum_{j=1}^{n} A_j(x)\frac{\partial}{\partial x_j} + A_o(x) ,$$

where $A_j \in C^\infty(U,\mathrm{End}(V))$. Let $f \in C^\infty(U)$ and $\varphi \in C^\infty(U,E|U)$. Then we get

$$D(f\varphi)(x) = f(x)D(\varphi)(x) + \Big(\sum_{j=1}^{n} \frac{\partial f}{\partial x_j}(x)A_j(x)\Big)\varphi(x) =$$

$$= \sigma_D(df_x,x)(\varphi(x)) + f(x)D(\varphi)(x) \quad . \quad \text{Q.E.D.}$$

Consider D as an operator acting on $L^2(X,E)$ with domain $C_c^\infty(X,E)$. According to Definition 5.3, D is symmetric. Therefore, there exist closed extensions of D. We observe that D has two natural closed extensions to unbounded operators in $L^2(X,E)$. The minimal extension is obtained by taking the closure of the graph of D. We say that $\varphi \in L^2(X,E)$ is in the minimal domain of D if there exists a sequence $\{\varphi_k\}_{k\in\mathbb{N}} \subset C_c^\infty(X,E)$ such that $\|\varphi_k-\varphi\| \longrightarrow 0$ for $k \longrightarrow \infty$, and $D\varphi_k$ converges to some element $\varphi' \in L^2(X,E)$. This limit $\varphi' = D\varphi$ is independent of the sequence chosen. The maximal extension of D is obtained by taking the domain to be all $\varphi \in L^2(X,E)$ such that the distributional image $D\varphi$ is in $L^2(X,E)$. In other words, $\varphi \in L^2(X,E)$ is in the maximal domain of D if the linear functional $\varphi' \longmapsto (\varphi,D\varphi')$ on

$C_c^\infty(X,E)$ is bounded in the L^2-norm. Now we have

PROPOSITION 5.7. Let $D : C_c^\infty(X,E) \longrightarrow C_c^\infty(X,E)$ be a generalized Dirac operator on a manifold X with a cusp of rank one. Then the minimal and the maximal extension of D coincide. In particular, D is essentially selfadjoint.

PROOF. We follow the proof of Theorem 1.17 in [37] . By definition, X is a complete Riemannian manifold. We choose a C^∞ function $h : \mathbb{R}^+ \longrightarrow [0,1]$ such that $h(t) = 1$ for $t \leq 1$, $h(t) = 0$ for $t \geq 2$ and $h'(t) \approx -1$ on $[1,2]$. Fix $x_o \in X$ and let $d : X \longrightarrow \mathbb{R}^+$ be a regularization of the function $\delta(x,x_o)$, δ the geodesic distance on X such that $\|\nabla d\| \leq 3/2$. For each $m \in \mathbb{N}$, we set

$$f_m(x) = h(\tfrac{1}{m}\, d(x)) \quad .$$

Then one has $\|\nabla f_m\| \leq 2/m$ and $\operatorname{supp}(f_m) \subset B(2m)$, where $B(r) = \{x \in X | d(x) \leq r\}$. The completeness of X implies that $B(r)$ is compact for all $r \in \mathbb{R}^+$. Let $\varphi \in L^2(X,E)$ be in the maximal domain of D . We have to show that φ is also in the minimal domain. Let $\varphi' \in C_c^\infty(X,E)$ and $f \in C^\infty(X)$. Employing Lemma 5.6. we get

$$(f\varphi, D\varphi') = (\varphi, D(\bar{f}\varphi')) - (\varphi, \sigma_D(d\bar{f})(\varphi')) =$$

$$= (\varphi, D(\bar{f}\varphi')) + (\sigma_D(df)\varphi, \varphi') \quad ,$$

where we used that $\sigma_D = -(\sigma_D)^*$. This shows that Lemma 5.6. continues to hold for $f \in C^\infty(X)$ and $\varphi \in L^2(X,E)$, if $D(f\varphi)$, $D\varphi$ are interpreted in the distributional sense. Now let

$$\varphi_m = f_m \varphi \in L^2(X,E) \quad .$$

The support of φ_m is compact and we have

$$D(\varphi_m) = \sigma_D(df_m)\varphi + f_m D(\varphi) \quad .$$

This shows that φ_m is in the maximal domain of D . By definition, D coincides on Y_1 with a first-order locally invariant operator $\mathcal{D} : C^\infty(Y,\mathcal{E}) \longrightarrow C^\infty(Y,\mathcal{E})$. Let $X_1,\dots,X_n$ be an orthonormal basis for p and V the space of the isotropy representations of $\tilde{\mathcal{E}}$. By Lemma 5.5. there exist $A_j \in \operatorname{End}(V)$, $j=1,\dots,n$, such that

$$\sigma_D(df_m) = \sum_{j=1}^{n} (X_j f_m) A_j \quad .$$

Now we observe that on G/K the Levi-Civita connection ∇ of the left invariant Riemannian metric of G/K coincides with the canonical left invariant connection [54, XI, Theorem 3.3]. Thus $\nabla_{X_j} f_m = R(X_j) f_m$. According to the definition of f_m, we have $f_m(x) = 1$ if $d(x) \le m$, i.e. $df_m = 0$ on $B(m)$. Since $\|\nabla f_m\| \le 2/m$ we obtain

$$\|\sigma_D(df_m)\varphi\| \le C\|\nabla f_m\|\,\|\varphi\| \le C_1 \frac{1}{m} \tag{5.8}$$

which implies that $\|D\varphi_m - D\varphi\| \to 0$ for $m \to \infty$. In view of this observation, it suffices to consider the case where both φ and $D\varphi$ are in L^2 and have compact support. Let $N = \partial Y_1$. Then we have $Y_1 \cong N \times [1,\infty)$. For $l \in \mathbb{N}$ we set $X_l = X_0 \cup (N \times [1, l+1])$. We have $X = X_l \cup Y_{l-1}$ where $Y_l = N \times [l,\infty)$. Choose $l \in \mathbb{N}$ such that $\operatorname{supp}\varphi$, $\operatorname{supp} D\varphi \subset X_{l-1}$ and let $\tilde{X} = X_l \cup (-X_l)$ be the double of X. There exists a vector bundle $\tilde{E}$ over $\tilde{X}$ which is the double of $E|X_l$.

Let $\pi : N \times [l, l+1] \to N \times \{l\}$ be the projection. The restriction of E to $N \times [l, l+1]$ is isomorphic to $\pi^*(E|N \times \{l\})$. Now we deform D on X to an elliptic operator $D' : C^\infty(X_l, E|X_l) \to C^\infty(X_l, E|X_l)$ which coincides with D on $X_{l-1} \cup (N \times [l, l+1/2])$ and which, in a neighborhood of ∂X_l, takes the special form

$$D' = \frac{\partial}{\partial r} + D_1$$

where $D_1 : C^\infty(N, E|N) \to C^\infty(N, E|N)$ is a first-order elliptic differential operator. Then there is a natural double of D'

$$\tilde{D} : C^\infty(\tilde{X}, \tilde{E}) \to C^\infty(\tilde{X}, \tilde{E}) \quad .$$

$\tilde{D}$ is an elliptic differential operator. Due to the compactness of $\tilde{X}$, the minimal and the maximal domain of $\tilde{D}$ coincide. This can be proved using a parametrix for $\tilde{D}$ (see [75]). Consider φ as an element of $L^2(\tilde{X}, \tilde{E})$. Since $\operatorname{supp}\varphi$, $\operatorname{supp} D\varphi \subset X_{l-1}$, it follows that $D\varphi = \tilde{D}\varphi$. Choose a sequence $\{\varphi_k\}_{k\in\mathbb{N}} \subset C^\infty(\tilde{X}, \tilde{E})$ such that $\varphi_k \to \varphi$ in $L^2(\tilde{X}, \tilde{E})$ and $\tilde{D}\varphi_k \to \tilde{D}\varphi$ in $L^2(\tilde{X}, \tilde{E})$ for $k \to \infty$. Let $g \in C_c^\infty(X_l)$ be such that $g \equiv 1$ on X_{l-1} and $g \equiv 0$ on $N \times [l+1/2, l+1] \subset X_l$. Since $\operatorname{supp}\varphi \subset X_{l-1}$, it is clear that $g\varphi_k \to \varphi$ in $L^2(X,E)$. Moreover, we have $\tilde{D}(g\varphi_k) = D(g\varphi_k)$ and, by Lemma 5.6,

$$\tilde{D}(g\varphi_k) = \sigma_{\tilde{D}}(dg)\varphi_k + g\tilde{D}(\varphi_k) \quad .$$

But $g\tilde{D}(\varphi_k) \to g\tilde{D}(\varphi)$ in $L^2(X,E)$ and $\sigma_{\tilde{D}}(dg)\varphi_k \to \sigma_{\tilde{D}}(dg)\varphi$ in $L^2(X,E)$. Since $\operatorname{supp} dg \cap \operatorname{supp}\varphi = \emptyset$, this implies that $D(g\varphi_k) \to D\varphi$ in $L^2(X,E)$, i.e. φ is in the minimal domain of D. Q.E.D.

Now introduce the operator

$$Q = D^2 \quad .$$

If we regard Q as an operator in the Hilbert space $L^2(X,E)$ with domain $C_c^\infty(X,E)$, then this operator is positive and symmetric and our next purpose is to show that Q acting in $L^2(X,E)$ with the domain above is essentially selfadjoint. To prove this fact we shall employ techniques similar to those used by Andreotti and Vesentini [3] .

As above, let $d : X \to \mathbb{R}^+$ be a regularization of the geodesic distance $\delta(x,x_o)$ from a fixed point $x_o \in X$ such that $\|\nabla d\| \leq 3/2$. For $B(r) = \{x \in X \mid d(x) \leq r\}$, $r \in \mathbb{R}^+$, and $\varphi \in C^\infty(X,E)$ we denote by $\|\varphi\|_{B(r)}$ the norm of the restriction of φ to $B(r)$. Recall that all $B(r)$, $r \in \mathbb{R}^+$, are compact.

LEMMA 5.9. There exists a constant $C > 0$ such that, for $0 < r < R$ and $\varphi \in C^\infty(X,E)$, one has the inequality

$$\|D\varphi\|^2_{B(r)} \leq \|Q\varphi\|^2_{B(R)} + \left(1+\frac{C}{(R-r)^2}\right)\|\varphi\|^2_{B(R)} \quad .$$

PROOF. The proof is similar to the proof of Proposition 3 in [3] . For the sake of completeness we give the details. Let $f : \mathbb{R} \to [0,1]$ be a C^∞-function such that $f(t) = 1$ for $t \leq 1$, $f(t) = 0$ for $t \geq 2$ and $f'(t) \approx -1$ on $[1,2]$. We introduce the function

$$g(x) = f\left(\frac{d(x)+R-2r}{R-r}\right)$$

for any choice of $R > r > 0$ with $d(x)$ as above. g is a C^∞-function with $0 \leq g(x) \leq 1$, $|\nabla g(x)| \leq 2/(R-r)$ and

$$g(x) = \begin{cases} 1 \ , & \text{if } x \in \overline{B(r)} \\ 0 \ , & \text{if } x \in X-B(R) \ . \end{cases}$$

If $\psi \in C^\infty(X,E)$ and the support of ψ is contained in $B(R)$, then we have

$$(D\varphi,D\psi) = (Q\varphi,\psi) \quad . \tag{5.10}$$

Here we used that D is formally selfadjoint. Now set $\psi = g^2\varphi$. Then, by Lemma 5.6, we get

$$D\psi = g^2 D\varphi + 2g\sigma_D(dg)\varphi \quad .$$

Substituting this in (5.10) we obtain

$$\|gD\varphi\|^2 \leq |(Q\varphi,g^2\varphi)_{B(R)}| + |(D\varphi,2g\sigma_D(dg)\varphi)_{B(R)}| \quad .$$

On the other hand, we have

$$|(Q\varphi, g^2\varphi)_{B(R)}| \leq \|Q\varphi\|_{B(R)} \|g^2\varphi\|_{B(R)} \leq \frac{1}{2}(\|Q\varphi\|^2_{B(R)} + \|g\varphi\|^2_{B(R)})$$

and similarly

$$|(D\varphi, 2g\sigma_D(dg)\varphi)_{B(R)}| \leq \frac{1}{2}(\|gD\varphi\|^2_{B(R)} + 4\|\sigma_D(dg)\varphi\|^2_{B(R)}) \quad .$$

This gives

$$\begin{aligned} \|gD\varphi\|^2_{B(R)} &\leq \|Q\varphi\|^2_{B(R)} + \|g\varphi\|^2_{B(R)} + 4\|\sigma_D(dg)\varphi\|^2_{B(R)} \leq \\ &\leq \|Q\varphi\|^2_{B(R)} + \|\varphi\|^2_{B(R)} + \frac{C}{(R-r)^2}\|\varphi\|^2_{B(R)} \quad , \end{aligned} \tag{5.11}$$

where $C > 0$ is a certain constant. To prove the second inequality we employed the estimation $|\nabla g(x)| \leq 2/(R-r)$ and the argument which let to (5.8) . Since $g \geq 0$ and $g \equiv 1$ on $B(r)$, we have $\|D\varphi\|_{B(r)} \leq \|gD\varphi\|_{B(R)}$, which together with (5.11) proves the Lemma. Q.E.D.

LEMMA 5.12. Let $\varphi \in C^\infty(X,E)$ and assume that $\|\varphi\| < \infty$ and $\|Q\varphi\| < \infty$. Then we have $\|D\varphi\| < \infty$ and

$$(Q\varphi, \varphi) = \|D\varphi\|^2 \quad .$$

PROOF. Let $\varphi \in C^\infty(X,E)$ be such that $\|\varphi\| < \infty$ and $\|Q\varphi\| < \infty$. Let $r > 0$ and apply Lemma 5.9 with $R=2r$. Letting r tend to ∞ we obtain $\|D\varphi\| < \infty$. This is the first part. Appealing to Proposition 5.7, it follows that there exists a sequence $\{\varphi_j\}_{j\in\mathbb{N}} \subset C^\infty_c(X,E)$ such that $\varphi_k \to \varphi$ in $L^2(X,E)$ and $D\varphi_k \to D\varphi$ in $L^2(X,E)$ for $k \to \infty$. This implies

$$(Q\varphi, \varphi) = \lim_{j\to\infty} (D^2\varphi, \varphi_j) = \lim_{j\to\infty} (D\varphi, D\varphi_j) = \|D\varphi\|^2 . \quad \text{Q.E.D.}$$

Using this lemma, we obtain

COROLLARY 5.13. Let $D : C^\infty_c(X,E) \to C^\infty_c(X,E)$ be a generalized Dirac operator. Put $Q = D^2$ and $\mathbb{D}(Q) = C^\infty_c(X,E)$. Then Q is essentially self-adjoint as an operator in the Hilbert space $L^2(X,E)$.

PROOF. Let $\lambda \in \mathbb{C} - \mathbb{R}$ and suppose that $\psi \in ((Q+\lambda I)C^\infty_c(X,E))^\perp$. Then $(\psi, (Q+\lambda I)\varphi) = 0$ for all $\varphi \in C^\infty_c(X,E)$, i.e. $(Q+\lambda I)\psi = 0$ in the distributional sense. Since Q is elliptic, the regularity theorem implies that $\psi \in C^\infty(X,E)$. Thus $\psi \in C^\infty(X,E) \cap L^2(X,E)$ and $Q\psi = -\lambda\psi$, i.e. $Q\psi \in L^2(X,E)$. Then Lemma 5.12 implies that $D\psi \in L^2(X,E)$ and

$$\lambda\|\psi\|^2 = -(Q\psi, \psi) = -\|D\psi\|^2 \quad .$$

Since $\lambda \notin \mathbb{R}$, this implies that $\psi = 0$. Thus $(Q+\lambda I)C_c^\infty(X,E)$ is dense in $L^2(X,E)$ for all $\lambda \in \mathbb{C} - \mathbb{R}$. But this is the well-known criterion for an operator in a Hilbert space to be essentially self-adjoint. Q.E.D.

Thus, the closure H of Q in L^2 is the unique selfadjoint extension of Q to an unbounded operator in $L^2(X,E)$. In particular, H coincides with Friedrichs' extension of Q , which shows that H is a positive self-adjoint operator.

We shall now introduce the Sobolev spaces. Let E be a vector bundle over X which is locally homogeneous at infinity and let $\mathcal{E}$ be the corresponding locally homogeneous vector bundle over Y . There is a unique G-invariant connection $\tilde{\nabla}$ on $\tilde{\mathcal{E}}$ such that, for $\varphi \in C^\infty(\tilde{Y},\tilde{\mathcal{E}})$ and $Z \in p$ one has

$$\tilde{\nabla}_{d\pi_e(Z)}(\varphi) = \frac{d}{dt}\varphi(\pi(\exp tZ))\Big|_{t=0} ,$$

where $\pi : G \to G/K = \tilde{Y}$ denotes the canonical projection. Being G-invariant, $\tilde{\nabla}$ can be pushed down to a connection $\nabla^{\mathcal{E}}$ on $\mathcal{E}$. Let ∇ be any connection on E which coincides on $E|Y_1 = \mathcal{E}|Y_1$ with the locally invariant connection $\nabla^{\mathcal{E}}$.

<u>DEFINTION 5.14</u>. Let $k \in \mathbb{N}$. The Sobolev space $H^k(X,E)$ is the space of all sections $\varphi \in L^2(X,E)$ such that $\nabla^l\varphi \in L^2(X, \overset{l}{\underset{1}{\otimes}} T^*X \otimes E)$ for $l \leq k$. The Sobolev norm of $\varphi \in H^k(X,E)$ is defined by

$$\|\varphi\|_k = \sum_{l=0}^{k} \|\nabla^l\varphi\| ,$$

where $\|\,.\,\|$ denotes the norm in $L^2(X,\overset{l}{\underset{1}{\otimes}} T^*X \otimes E)$.

We observe that different choices of connections ∇ , which coincide on Y_1 with $\nabla^{\mathcal{E}}$, lead to equivalent norms on $H^k(X,E)$.

If $k=1$ or $k=2$, then one can use any generalized Dirac operator to introduce equivalent norms.

<u>PROPOSITION 5.15</u>. Let $D : C_c^\infty(X,E) \to C_c^\infty(X,E)$ be a generalized Dirac operator. The norm of the Sobolev space $H^2(X,E)$ is equivalent to the norm

$$|||\varphi|||_2 = \|(I+D^2)\varphi\| , \quad \varphi \in H^2(X,E) ,$$

and the norm of the Sobolev space $H^1(X,E)$ is equivalent to the norm

$$(\|D\varphi\|^2 + \|\varphi\|^2)^{1/2} \quad , \qquad \varphi \in H^1(X,E) \quad .$$

Here $D^2\varphi$ and $D\varphi$ are interpreted in the distributional sense. Moreover, $C_c^\infty(X,E)$ is a dense subspace of $H^k(X,E)$, for $k=1,2$.

PROOF. For $l \in \mathbb{N}$, let $X_l = X-Y_{l+1}$, where Y_{l+1} is given by (2.13) . Let $\widetilde{X} = X_2 \cup (-X_2)$ and $\widetilde{E} = E|X_2 \cup E|X_2$ be the natural doubles. As in the proof of Proposition 5.7, we construct an elliptic first-order differential operator $\widetilde{D} : C^\infty(\widetilde{X},\widetilde{E}) \to C^\infty(\widetilde{X},\widetilde{E})$ which coincides with D on X_1 . Let $\widetilde{H} = \widetilde{D}^2$. Then $\widetilde{H}$ is a positive elliptic differential operator which coincides with D^2 on X_1 . Moreover, let ∇_o be a connection of $\widetilde{E}$ which coincides with ∇ on X_1 . Choose $f \in C^\infty(X)$ such that $f \equiv 0$ on X_o and $f \equiv 1$ on Y_2 . Set $h = 1-f$ and let $\varphi \in H^2(X,E)$. Using the formula $\nabla(h\varphi) = dh \otimes \varphi + h\nabla\varphi$, it is easy to see that $(\nabla_o)^l(h\varphi) \in L^2(\widetilde{X},\widetilde{E})$ for $l \leq 2$. Since $\widetilde{X}$ is a closed manifold, this implies that $h\varphi \in H^2(\widetilde{X},\widetilde{E})$, where $H^2(\widetilde{X},\widetilde{E})$ denotes the usual Sobolev space for a compact manifold (see [11, p.511]) . $I+\widetilde{H}$ is a positive definite second-order elliptic differential operator. It is a standard result for compact manifolds that in this case the Sobolev norm in $H^2(\widetilde{X},\widetilde{E})$ is equivalent to $\|(I+\widetilde{H})\varphi\|$. But $\widetilde{H}(h\varphi) = D^2(h\varphi)$. Therefore $(I+D^2)(h\varphi) \in L^2(X,E)$ and there exist constants $C_1 > 0$ and $C_2 > 0$ such that

$$C_1 \|(I+D^2)(h\varphi)\| \leq \|h\varphi\|_2 \leq C_2\|(I+D^2)(h\varphi)\| \quad , \tag{5.16}$$

for all $\varphi \in H^2(X,E)$.

Now we consider $f\varphi$. Let $\mathcal{D} : C^\infty(Y,\mathcal{E}) \to C^\infty(Y,\mathcal{E})$ be the locally invariant operator associated to D and let V be the space of the isotropy representation of $\widetilde{E}$. By definition, we have $\operatorname{supp} f\varphi \subset Y_1$. Thus $f\varphi \in L^2(Y,\mathcal{E})$. Let $\nabla^{\mathcal{E}}$ be the connection of $\mathcal{E}$ introduced above. By definition, we have $\nabla(f\varphi) = \nabla^{\mathcal{E}}(f\varphi)$. Using the formula $\nabla(f\varphi) = df \otimes \varphi + f\nabla\varphi$, it follows that $(\nabla^{\mathcal{E}})^l(f\varphi) \in L^2(Y,\mathcal{E})$ for $l \leq 2$. Let $H^2(Y,\mathcal{E})$ be the space of all $\psi \in L^2(Y,\mathcal{E})$ such that $(\nabla^{\mathcal{E}})^l\psi \in L^2(Y,\mathcal{E})$, for $l \leq 2$, the derivatives being understood in the distributional sense. In order to complete the proof of the first statement, it is sufficient to show that for each $\psi \in H^2(Y,\mathcal{E})$ one has $(I+\mathcal{D}^2)\psi \in L^2(Y,\mathcal{E})$ and there exist constants $C_3>0$ and $C_4>0$ such that

$$C_3 \|(I+\mathcal{D}^2)\psi)\| \leq \sum_{l=0}^{2} \|(\nabla^{\mathcal{E}})^l\psi\| \leq C_4 \|(I+\mathcal{D}^2)\psi\|. \tag{5.17}$$

For simplicity we shall drop the index $\mathcal{E}$ of $\nabla^{\mathcal{E}}$. Using (5.4) combined with (3.3), we obtain

$$\tilde{\mathcal{D}}^2 = \tilde{\nabla}^*\tilde{\nabla} + Q(\mathrm{Id} \otimes L_3)Q , \tag{5.18}$$

for a certain $L_3 \in \mathrm{End}(V)$. Q denotes the projection (3.2) . This implies the left inequality in (5.17). In order to obtain the other inequality, we proceed as in [24] . First we observe that the operator $\mathcal{D}^2 : C_c^\infty(Y,E) \longrightarrow C_c^\infty(Y,E)$, considered as an operator in $L^2(Y,E)$, is essentially self-adjoint. This is a consequence of Corollary 1.2 in [61] . Now let $\psi \in H^2(Y,E)$. By the first inequality in (5.17), we have $(I+\mathcal{D}^2)\psi \in L^2(Y,E)$. In view of the essential self-adjointness of $\mathcal{D}^2$ there exists a sequence $\{\psi_j\}_{j\in\mathbb{N}} \subset C_c^\infty(Y,E)$ such that $(I+\mathcal{D}^2)\psi_j \longrightarrow (I+\mathcal{D}^2)\psi$ in $L^2(Y,E)$ for $j\to\infty$. Therefore, in order to prove the second inequality, we may assume that $\psi \in C_c^\infty(Y,E)$. We now follow the proof of Theorem 1.3 in [24]. First of all, we have

$$\|\nabla\psi\|^2 = (\nabla^*\nabla\psi,\psi) \leq \frac{1}{2}(\|\nabla^*\nabla\psi\| + \|\psi\|)^2 . \tag{5.19}$$

In view of (5.18) this implies

$$\|\nabla\psi\| \leq C(\|\psi\| + \|\mathcal{D}^2\psi\|) .$$

The Bianchi identity gives

$$\nabla^*\nabla - \nabla\nabla^* = R , \tag{5.20}$$

where R denotes an algebraic operator defined in terms of the curvature tensor of G/K . Using (5.19) and (5.20), we get

$$\|\nabla^2\psi\|^2 = (\nabla^*\nabla^2\psi,\nabla\psi) = (\nabla\nabla^*\nabla\psi,\nabla\psi) + (R\nabla\psi,\nabla\psi) \leq$$

$$\leq \|\nabla^*\nabla\psi\|^2 + C\|\nabla\psi\| \leq C_1(\|\nabla^*\nabla\psi\|^2)$$

which together with (5.18) and (5.19) gives

$$\sum_{l=0}^{2} \|\nabla^l\psi\|^2 \leq C_4\|(I+\mathcal{D}^2)\psi\|, \qquad \psi \in C_c^\infty(Y,E) . \tag{5.21}$$

In view of the essential self-adjointness and the left inequality, which we already proved, (5.21) can be extended to all $\psi \in H^2(Y,E)$. This proves the first statement of the proposition. The proof of the second statement runs along the same lines. Let $\varphi \in H^1(X,E)$. Repeating the arguments given above, it follows that $(I+\tilde{H})^{1/2}(h\varphi) \in L^2(\tilde{X},\tilde{E})$ and there exist constants $C_1 > 0$ and $C_2 > 0$ such that

$$C_1\|(I+\tilde{H})^{1/2}(h\varphi)\| \leq \|h\varphi\|_1 \leq C_2\|(I+\tilde{H})^{1/2}(h\varphi)\| ,$$

$\varphi \in H^1(X,E)$. Recall that $\tilde{H}(h\varphi) = D^2(h\varphi)$. Thus, in view of Lemma 5.12, we get

$$\|(I+\tilde{H})^{1/2}(h\varphi)\| = ((I+D^2)(h\varphi),h\varphi) = \|D(h\varphi)\|^2 + \|h\varphi\|^2 .$$

Furthermore, using Lemma 5.12 and (5.18), we obtain

$$C_3(\|D(f\varphi)\|^2 + \|f\varphi\|^2) \leq ((\nabla^*\nabla+I)(f\varphi),f\varphi) \leq$$

$$\leq C_4(\|D(f\varphi)\|^2 + \|f\varphi\|^2)$$

for certain constants $C_3 > 0$ and $C_4 > 0$ and all $\varphi \in H^1(X,E)$. Combining these results we obtain the second statement.

Consider $T = I+D^2$ as an operator in $L^2(X,E)$ with $\mathbb{D}(T) = C_c^\infty(X,E)$. By corollary 5.13, T is essentially self-adjoint, i.e. $\overline{T} = \overline{T}^* = T^*$. According to the definition, the domain of T^* consists of all $\varphi \in L^2(X,E)$ such that $T\varphi \in L^2(X,E)$. As we have seen above, this is precisely the Sobolev space $H^2(X,E)$. On the other hand, the domain of $\overline{T}$ is the closure of $C_c^\infty(X,E)$ in $H^2(X,E)$. This shows that $C_c^\infty(X,E)$ is dense in $H^2(X,E)$. Finally, by Proposition 5.7, the minimal and the maximal extensions of D coincide. According to the second statement of our proposition, the domain of the maximal extension coincides with $H^1(X,E)$ and the domain of the minimal extension is the closure of $C_c^\infty(X,E)$ in $H^1(X,E)$. Thus, $C_c^\infty(X,E)$ is dense in $H^1(X,E)$. Q.E.D.

COROLLARY 5.22. Let $\overline{D}$ and H be the closure in L^2 of D and D^2 , respectively. The domain of H is the Sobolev space $H^2(X,E)$ and the domain of $\overline{D}$ is the Sobolev space $H^1(X,E)$. Furthermore, consider the quadratic form $q(\varphi) = (D^2\varphi,\varphi)$, $\varphi \in C_c^\infty(X,E)$. The domain of the closure of q is $\mathbb{D}[H] = H^1(X,E)$.

The proof is an immediate consequence of Proposition 5.15.

CHAPTER VI
THE SPECTRAL RESOLUTION OF H

Let $D : C_c^\infty(X,E) \to C_c^\infty(X,E)$ be a generalized Dirac operator and H the unique selfadjoint extension of D^2 acting in $L^2(X,E)$. In this chapter we shall investigate the spectral resolution of H. First we consider the absolutely continuous part H_{ac} of H. To study H_{ac} we introduce a "free Hamiltonian" H_o which is obtained from H by imposing additional Neumann boundary conditions along ∂Y_1. Then we apply techniques from scattering theory to show that the wave operators $W_\pm(H,H_o)$ exist and are complete. This implies that H_{ac} is unitarily equivalent to the absolutely continuous part $H_{o,ac}$ of H_o. This fact combined with Theorem 4.38 leads to the complete description of H_{ac}.

Our approach to prove the existence and completeness of the wave operators uses the method of Enss [31]. We shall apply the abstract version of this approach to scattering theory developed by Amrein, Pearson and Wollenberg [2], [16, IV, §15].

To begin with we introduce some notation. For all details we refer to [16, IV, §15]. Let $\mathcal{H}$ be a separable Hilbert space and let H and H_o be two selfadjoint operators in $\mathcal{H}$. By P^{ac} (resp. P_o^{ac}) we shall denote the orthogonal projection of $\mathcal{H}$ onto the absolutely continuous subspace $\mathcal{H}_H^{ac}$ (resp. $\mathcal{H}_{H_o}^{ac}$) of H (resp. H_o). The wave operators $W_\pm(H,H_o)$ are defined by

$$W_\pm(H,H_o) = \operatorname*{s-lim}_{t\to\pm\infty} e^{itH} e^{-itH_o} P_o^{ac} \quad , \tag{6.1}$$

if this limit exists. The wave operators are called complete if $W_\pm = W_\pm(H,H_o)$ is an isometry of $\mathcal{H}_{H_o}^{ac}$ onto $\mathcal{H}_H^{ac}$.

Moreover, one has

$$HW_\pm \supset W_\pm H_o \ ,$$

i.e., if the wave operators exist and are complete then $H \restriction \mathcal{H}_H^{ac}$ is unitary equivalent to $H_o \restriction \mathcal{H}_{H_o}^{ac}$.

Let $C_\infty(\mathbb{R})$ be the space of all continuous functions on $\mathbb{R}$

vanishing at infinity. For any closed countable subset $I \subset \mathbb{R}$, we denote by $C_\infty(\mathbb{R}-I)$ the set of all functions $f \in C_\infty(\mathbb{R})$ satisfying $f(x) = 0$ for $x \in I$. A subset A_I of the space $C(\mathbb{R})$ of all bounded continuous functions on $\mathbb{R}$ is called multiplicative-generating for $C_\infty(\mathbb{R}-I)$, if the linear span of the set $\{f \mid f = hg,\ h \in A_I,\ g \in C_0^\infty(\mathbb{R}-I)\}$ is dense in $C_\infty(\mathbb{R}-I)$ with respect to the norm $\|f\| = \sup_{x\in\mathbb{R}} |f(x)|$.

Now we can state a version of the main result of [2] in the form we shall use it.

THEOREM 6.2. Let H and H_0 be two selfadjoint operators in a Hilbert space $\mathcal{H}$ and let $R_H(\zeta)$ (resp. $R_{H_0}(\zeta)$) denote the resolvent of H (resp. H_0). Assume that there exist selfatjoint operators P_+, P_- in $\mathcal{H}$ and a set A_I of multiplicative-generating functions satisfying the following properties

(i) $P_0^{ac} = P_+ + P_-$ and

$$\underset{t\to\pm\infty}{\text{s-lim}}\ e^{itH_0} P_\mp e^{-itH_0} P_0^{ac} = 0 \ .$$

(ii) $(I-P_0^{ac})\alpha(H_0)$ is compact for all $\alpha \in A_I$.

(iii) $R_H(i) - R_{H_0}(i)$ is compact.

(iv) $\int_0^{\pm\infty} \|(R_H(i) - R_{H_0}(i))e^{-itH_0}\alpha(H_0)P_\pm\|\, dt < \infty$

for all $\alpha \in A_I$.

Then the wave operators $W_\pm(H,H_0)$ exist and are complete. Moreover, H and H_0 have no singularly continuous spectrum and each eigenvalue of H and H_0 in $\mathbb{R} - I$ is of finite multiplicity. These eigenvalues accumulate at most at points of $I \cup \{\pm\infty\}$.

This is Corollary 19 in [16,IV,§15]. The proof is given there.

We shall apply this theorem to our situation. $\mathcal{H}$ is now the Hilbert space $L^2(X,E)$ and H is the unique selfadjoint extension of D^2. To define H_0, consider the decomposition $X = X_1 \cup Y_2$ as in Ch.V and let E_1 (resp. E_2) denote the restriction of E to X_1 (resp. Y_2). Let $H^1(Y_2,E_2)$ be defined as in (4.3) and set

$$H^1(X_1,E_1) = \{\varphi | X_1 \mid \varphi \in H^1(X,E)\} \ .$$

Consider the quadratic form

$$q(\varphi) = ||D\varphi||^2$$

acting on $H^1(X_1,E_1) \oplus H^1(Y_2,E_2)$ and let H_o be the associated self-adjoint operator on $L^2(X,E)$. H_o splits in the direct sum of an operator H_1 acting on $L^2(X_1,E_1)$ and an operator H_2 acting on $L^2(Y_2,E_2)$. Since X_1 is a compact manifold, it follows that $(H_1+I)^{-1}$ is a compact operator. Moreover, in view of Definition 5.3, H_2 satisfies the assumptions of Ch.4. Thus the spectral resolution of H_2 is described by Theorem 4.38. Let $\mathcal{E}_M$ be the locally homogeneous vector bundle over $\Gamma_M\backslash X_M$ associated to E_2 and let

$$\Delta_M : C^\infty(\Gamma_M\backslash X_M,\mathcal{E}_M) \longrightarrow C^\infty(\Gamma_M\backslash X_M,\mathcal{E}_M)$$

be the differential operator which corresponds to H_2. The first observation is

<u>LEMMA 6.3</u>. $R_H(-1) - R_{H_o}(-1)$ is a compact operator.

<u>PROOF</u>. Let $f \in C^\infty(X)$ be such that $f=0$ on X_2 and $f=1$ in the complement of a compact set. Then we may write

$$R_H(-1) - R_{H_o}(-1) = (f-1)R_{H_o}(-1) - R_H(-1)((H+I)fR_{H_o}(-1)-I) .$$

First consider $(f-1)R_{H_o}(-1)$. Let $L^2_o(Y_2,E_2)$ be the space of cusp forms in $L^2(Y_2,E_2)$. It follows from the proof of Proposition 4.9 that $R_{H_o}(-1)$ is a compact operator on $L^2(X_1,E_1) \oplus L^2_o(Y_2,E_2)$. According to Ch. IV, the orthogonal complement to this space in $L^2(X,E)$ can be identified with $L^2([c,\infty) \times \Gamma_M\backslash X_M,\mathcal{E}_M)$ and H_o acts on this space via

$$-y^2 \frac{\partial^2}{\partial y^2} + (m-1)y\frac{\partial}{\partial y} + \Delta_M .$$

Let $\tilde{\mu}_1 < \tilde{\mu}_2 < \ldots$ be the eigenvalues of Δ_M and $E(\tilde{\mu}_j)$ the eigenspace corresponding to the eigenvalue $\tilde{\mu}_j$. Given $k \in \mathbb{N}$, denote by P_k the orthogonal projection of $L^2([c,\infty) \times \Gamma_M\backslash X_M,\mathcal{E}_M)$ onto the subspace $L^2([c,\infty)) \otimes (\bigoplus_{j=1}^{k} E(\tilde{\mu}_j))$. Denote by $\tilde{R}_{H_o}(-1)$ the restriction of $R_{H_o}(-1)$ to $L^2([c,\infty) \times \Gamma_M\backslash X_M,\mathcal{E}_M)$. Then $\tilde{R}_{H_o}(-1)P_K$ converges strongly to $\tilde{R}_{H_o}(-1)$ as $k \longrightarrow \infty$. On the other hand, it is easy to see that $(f-1)\tilde{R}_{H_o}(-1)P_k$ is a compact operator. Since the compact operators form a closed subspace in the space of linear operators on $L^2(X,E)$ with respect to the strong operator topology, it follows that $(f-1)\tilde{R}_{H_o}(-1)$ is compact. Now observe that, on Y_2, we have $D^2 = \nabla^*\nabla + L$ where L is an operator of order zero. This implies

$$(H + I)(fR_{H_o}(-1)) = (f - 1) + 2\nabla f\cdot\nabla R_{H_o}(-1) + \Delta f\cdot R_{H_o}(-1) .$$

As above, it follows that $\nabla f.\nabla R_{H_o}(-1)$ and $\nabla f.R_{H_o}(-1)$ are compact operators. Similar arguments show that $R_H(-1)(f-1)$ is compact too. Q.E.D.

To be able to apply Theorem 6.2 in our case, we have to construct a decomposition $P_o^{ac} = P_+ + P_-$ and to select a family A_I of multiplicative-generating functions which satisfy the assumptions of Theorem 6.2.

Let $\mu_1 \le \mu_2 \le \ldots \longrightarrow \infty$ be the eigenvalues of Δ_M, counted to multiplicity and set

$$I = \{\frac{m^2}{4} + \mu_k \mid k \in \mathbb{N}\} , \tag{6.4}$$

where m is defined in Theorem 4.38. We put $A_I = C_o^\infty(\mathbb{R}-I)$. Then it is clear that A_I is multiplicative-generating for $C_\infty(\mathbb{R}-I)$. Now we turn to the construction of the operators P_+ and P_- . Using the observations made before Lemma 6.3, it follows from Theorem 4.38 that the absolutely continuous part $H_{o,ac}$ of H_o can be identified with the operator (4.29) acting on $L^2([c,\infty)\times \Gamma_M\backslash X_M, E_M)$, $c = 2^{1/|\lambda|}$. We use the spectral resolution V^*J of T_ν given by (4.33) and (4.36) to construct $P_\pm$ on this space. Let

$$\bigoplus_{k=1}^{\infty} L^2([\frac{m^2}{4} + \mu_k, \infty)) \longrightarrow L^2(\mathbb{R};\ell^2)$$

be the embedding obtained by extending functions by zero, and let Z_+ be the orthogonal projection of $L^2(\mathbb{R};\ell^2)$ onto this subspace. The Fourier transformation defines a unitary operator F_o in $L^2(\mathbb{R};\ell^2)$. Furthermore, let K_o and K_1 be the selfadjoint operators in $L^2(\mathbb{R};\ell^2)$ which are given by $(K_o u)(x) = xu(x)$ and $(K_1 u)(x) = -iu'(x)$, $u \in C_o^\infty(\mathbb{R};\ell^2)$. Then we have $F_o^* K_o = K_1 F_o^*$ and $Z_+ K_o = BZ_+$, where B is defined by (4.37). Let χ_+ and χ_- denote the multiplication operators by the characteristic functions $\chi_{[0,\infty)}$ and $\chi_{(-\infty,0]}$, respectively, and set

$$\tilde{P}_\pm = Z_+ F_o \chi_\pm F_o^* Z_+^* . \tag{6.5}$$

It is clear that $\tilde{P}_\pm$ are self-adjoint operators in the Hilbert space

$$\bigoplus_{k=1}^{\infty} L^2([\frac{m^2}{4} + \mu_k, \infty)) .$$

Moreover, we have

$$\tilde{P}_{\pm}e^{-itB} = Z_+ F_o \chi_{\pm} e^{-itK_1} F_o^* Z_+^* . \tag{6.6}$$

Let $u \in L^2(\mathbb{R};\ell^2)$. Using the Fourier transformation, it follows that $(e^{-itK_1}u)(x) = u(x-t)$. Thus we get

$$\| e^{itK_1}\chi_{\pm} e^{-itK_1}u\|^2 = \pm \int_{-t}^{\pm\infty} \|u(x)\|^2_{\ell^2}dx \longrightarrow 0$$

as $t \to \mp\infty$, which together with (6.6) implies that

$$\underset{t\to\pm\infty}{\text{s-lim}}\ e^{itB}\,\tilde{P}_{\mp}\, e^{-itB} = 0 .$$

Finally, let

$$P_{\pm} = J^*V\tilde{P}_{\pm}V^*J , \tag{6.7}$$

where J and V are the isometries (4.33) and (4.36), respectively. $P_{\pm}$ are selfadjoint operators on $H^{ac}_{H_o}$. On the orthogonal complement of $H^{ac}_{H_o}$ we set $P_{\pm} = 0$. In this way we obtain self-adjoint operators on $H = L^2(X,E)$ satisfying

$$\begin{aligned} &P_o^{ac} = P_+ + P_- \quad \text{and} \\ &\underset{t\to\pm\infty}{\text{s-lim}}\ e^{itH_o} P_{\mp} e^{-itH_o} = 0 . \end{aligned} \tag{6.8}$$

This is condition (i) of Theorem 6.2. Next we shall verify condition (iv) . Given $t > 0$, let χ_t be the characteristic function of Y_{e^t} in X . Since H and H_o are positive operators, we can replace the point i by -1 . Let $\delta > 0$. Then we have

$$\begin{aligned} &\|(R_H(-1) - R_{H_o}(-1))\exp(\mp itH_o)\alpha(H_o)P_{\pm}\| \le \\ &\le \|R_H(-1) - R_{H_o}(-1)\|\ \|(1-\chi_{\delta t})\exp(\mp itH_o)\alpha(H_o)P_{\pm}\| + \\ &\|(R_H(-1) - R_{H_o}(-1))\chi_{\delta t}\|\ \|\alpha(H_o)\| . \end{aligned} \tag{6.9}$$

We shall prove that for each $\alpha \in C_o^\infty(\mathbb{R}-I)$ there exists $\delta > 0$ such that the right hand side is an integrable function of $t \in \mathbb{R}^+$. To estimate the first term on the right hand side we need the following auxiliary result :

<u>LEMMA 6.10</u>. Let $a \in \mathbb{R}$ and $f \in C_o^\infty(\mathbb{R}-\{a\})$. Choose $\varepsilon > 0$ such that $f(\lambda^2+a) = 0$ for $|\lambda| < \varepsilon$. Then, for every $m \in \mathbb{N}$, there exists a constant C which depends on f and m , such that, for $t \in \mathbb{R} - \{0\}$

and $|y| < \frac{\varepsilon}{2}|t|$, one has

$$|\int_0^\infty e^{2iy\lambda+it\lambda^2} f(\lambda^2+a)\,d\lambda| \le C|t|^{-m} .$$

PROOF. Let $t \neq 0$ and set $x = y/t$. The left hand side of the inequality equals

$$|\int_0^\infty e^{it(\lambda+x)^2} f(\lambda^2+a)\,d\lambda| = (2t)^{-m}|\int_0^\infty e^{it(\lambda+x)^2}\left(\frac{d}{d\lambda}\left(\frac{1}{\lambda+x}\right)\right)^m f(\lambda^2+a)d\lambda| \quad (6.11)$$

Now assume that $|y| < \varepsilon|t|/2$. Then $|x| < \varepsilon/2$. On the other hand, we have $f(\lambda^2+a) = 0$ for $|\lambda| < \varepsilon$. Thus, if $f(\lambda^2+a) \neq 0$ then we have $|\lambda+x| \geq |\lambda| - |x| > \varepsilon/2$. Hence (6.11) can be estimated by $C|t|^{-m}$. Q.E.D.

Let $\{\Phi_j\}_{j\in\mathbb{N}} \subset L^2(\Gamma_M\backslash X_M, E_M)$ be an orthonormal basis of eigenfunctions of Δ_M with eigenvalues $\{\mu_j\}_{j\in\mathbb{N}}$. Let $\varphi \in L^2([c,\infty)\times\Gamma_M\backslash X_M, E_M)$. If we use (4.34), (4.35) and the construction of P_+, then we get

$$(e^{-itH_o}\alpha(H_o)\varphi)(r,x) = \quad (6.12)$$

$$= \sum_{k=1}^\infty \frac{1}{2\pi}\left(\int_0^\infty \eta(r,\tfrac{m}{2} - i\zeta)\, e^{-it(\zeta^2+m^2/4+\mu_k)}\alpha(\zeta^2 + m^2/4 + \mu_k)\cdot\right.$$

$$\left.\cdot(J\varphi)(\zeta,k)d\zeta\right)\Phi_k(x) .$$

Let $\varphi = P_+v$, $v \in L^2([c,\infty)\times\Gamma_M\backslash X_M, E_M)$ and put $w = F_o^*Z_o^*V^*Jv$. Then $w \in L^2(\mathbb{R};\ell^2)$ and $J\varphi = VZ_+F_o(\chi_+w)$. If we use the definition of V, Z_+ and F_o, then we get

$$(J\varphi)(\zeta,k) = \sqrt{2\zeta}\lim_{X\to\infty}\int_0^X e^{-i(\zeta^2+m^2/4+\mu_k)s}\, w(s,k)\,ds ,$$

where the limit is taken in L^2. Assume that $t > 0$. If we insert this expression in (6.12) and then switch the order of integration, we obtain

$$(e^{-itH_o}\alpha(H_o)P_+v)(r,x) = \quad (6.13)$$

$$= \frac{1}{\sqrt{2}\,\pi}\sum_{k=1}^\infty \left(\int_0^\infty w(s,k)\int_0^\infty \eta(r,\tfrac{m}{2} - i\zeta)\, e^{-i(t+s)(\zeta^2+m^2/4+\mu_k)}\cdot\right.$$

$$\left.\cdot\,\alpha(\zeta^2+m^2/4+\mu_k)d\zeta\, ds\right)\Phi_k(x) .$$

Moreover, there exists $\varepsilon > 0$ such that $\alpha(\zeta^2 + m^2/4 + \mu_k) = 0$ for $|\zeta| < \varepsilon$ and all $k \in \mathbb{N}$. Assume that $|\log(y)| < \varepsilon t/2$. If we apply Lemma 6.10 to the interior integral on the right hand side of (6.13), then it follows that, for each $m \geq 1$, there exists a constant $C > 0$ such that

$$\int_{\Gamma_M \backslash X_M} |e^{-itH_o}\alpha(H_o)P_+ v(r,x)|^2 \, dx \leq$$

$$\leq C \sum_{k=1}^{\infty} \int_0^{\infty} |w(s,k)|^2 \int_0^{\infty} (t+s)^{-2m} ds = \frac{C}{2m} \|w\|^2 \, t^{-2m+1} \leq C_1 \|v\|^2 \, t^{-2m+1}.$$

Thus, for every $\alpha \in C_o^{\infty}(\mathbb{R} - I)$ there exist $C > 0$ and $\delta > 0$ such that, for $t > \delta^{-1}$, one has

$$\|(1 - \chi_{\delta t})e^{-itH_o}\alpha(H_o)P_+\| \leq C\, t^{-2} .$$

In the same manner one can show that

$$\|(1 - \chi_{\delta t})e^{itH_o}\alpha(H_o)P_-\| \leq C\, t^{-2} , \quad \text{for} \quad t > \delta^{-1} .$$

Hence, for this choice of δ, the first term on the right hand side of (6.9) is integrable as a function of $t \in \mathbb{R}^+$.

Now consider the second term on the right hand side of (6.9). As in the proof of Lemma 6.3 we obtain

$$(R_H(-1) - R_{H_o}(-1))\chi_{\delta t} = (f - 1)R_{H_o}(-1)\chi_{\delta t} - R_H(-1)(2\nabla f \cdot \nabla R_{H_o}(-1)\chi_{\delta t} + \Delta f \cdot R_{H_o}(-1)\chi_{\delta t}) \tag{6.14}$$

for $t \gg 0$. Now observe that $R_{H_o}(-1)\chi_{\delta t}$ acts on $L^2(Y_2, E_2)$. Let $\varphi \in L_o^2(Y_2, E_2)$. Then $R_{H_o}(-1)\varphi \in L_o^2(Y_2, E_2)$ and in view of (4.12), we obtain

$$\|R_{H_o}(-1)\chi_{\delta t}\varphi\| = \|\chi_{\delta t}R_{H_o}(-1)\varphi\| \leq Ce^{-2\delta t}\|R_{H_o}(-1)\varphi\|_1 \leq \leq C_1 e^{-2\delta t}\|\varphi\| . \tag{6.15}$$

Using the results of Ch.4, it follows that, on the orthogonal complement of $L_o^2(Y_2, E_2)$ in $L^2(Y_2, E_2)$, $R_{H_o}(-1)$ is given by the kernel

$$\sum_{j=1}^{\infty} g_j(r,r')\,\Phi_j(x) \otimes \Phi_j(x')$$

where

$$g_j(r,r') = \frac{(rr')^{m/2}}{\sqrt{m^2/4 + \mu_j + 1}} \begin{cases} (r/r')^{(m^2/4 + \mu_j + 1)^{1/2}} & , \quad r' > r \\ (r'/r)^{(m^2/4 + \mu_j + 1)^{1/2}} & , \quad r > r' \end{cases} .$$

Let $h \in C_o^\infty(X)$. Given $\varphi \in (L_o^2(Y_2,E_2))^\perp$, it follows that

$$\| h\cdot R_{H_o}(-1)\chi_{\delta t}\varphi\| \leq C \sum_{j=1}^{\infty} \exp(-\delta t(m^2/4 + \mu_j + 1)^{1/2})\|\varphi\| . \qquad (6.16)$$

Combining (6.15) and (6.16), we obtain

$$\| h\cdot R_{H_o}(-1)\chi_{\delta t}\| \leq C_2 e^{-\delta t} .$$

In view of (6.14), this shows that $\|(R_H(-1) - R_{H_o}(-1))\chi_{\delta t}\|$ is an integrable function of t. Thus we have verified condition (iv) of Theorem 6.2. Condition (iii) is a consequence of Lemma 6.3. It remains to establish condition (ii). Observe that $I - P_o^{ac}$ is the orthogonal projection of $L^2(X,E)$ onto the subspace which is spanned by the eigenfunctions of H_o. Using Proposition 4.39 and the fact that X_1 is compact, it follows that the eigenvalues of H_o have no finite point of accumulation. Thus $(I - P_o^{ac})\alpha(H_o)$ has finite range for each function $\alpha \in C_o^\infty(\mathbb{R} - I)$. Therefore, we have verified that for our choice of $\mathcal{H}$, H and H_o the assumptions of Theorem 6.2 are satisfied and we can apply Theorem 6.2 in our situation. This leads to

THEOREM 6.17. Let H and H_o be as above. Then

(i) The wave operators $W_\pm(H,H_o)$ exist and intertwine the absolutely continuous parts H_{ac} and $H_{o,ac}$ of H and H_o, respectively.

(ii) H has no singularly continuous spectrum.

As observed above, H_o splits in the direct sum of H_1 acting in $L^2(X_1,E_1)$ and H_2 acting in $L^2(Y_2,E_2)$. H_1 has pure point spectrum and the absolutely continuous part of H_2 is described by Theorem 4.38.

This gives

<u>COROLLARY 6.18</u>. Let $D : C_c^\infty(X,E) \longrightarrow C_c^\infty(X,E)$ be a generalized Dirac operator and H the unique selfadjoint extension of D^2. Further, let $\Delta_M : C^\infty(\Gamma_M\backslash X_M, E_M) \longrightarrow C^\infty(\Gamma_M\backslash X_M, E_M)$ be the differential operator associated to D^2 and let T_ν be the selfadjoint operator acting on the Hilbert space $L^2([c,\infty) \times \Gamma_M\backslash X_M, E_M)$ which is obtained from

$$T = -r^2 \frac{\partial^2}{\partial r^2} + (m-1)r\frac{\partial}{\partial r} + \Delta_M$$

by imposing Neumann boundary conditions at r=c. Then H_{ac} is unitarily equivalent to the absolutely continuous part of T_ν .

<u>REMARK 6.19</u>.

(i) At the beginning of Ch.V we observed that $d + \delta : \Lambda^*(X) \longrightarrow \Lambda^*(X)$ is a generalized Dirac operator. Thus Theorem 6.17 and Corollary 6.18 can be applied to the Laplacian $\Delta = (d+\delta)^2$ acting on $\Lambda^*(X)$. Since the space $\Lambda^p(X)$ of differential forms of degree p, $0 \leq p \leq n$, is invariant under Δ , we get a complete description of the continuous spectrum of the Laplacian Δ_p acting on $\Lambda^p(X)$. Compared with the stationary approach used by Faddejev (c.f. [55]), Enss' method is very effective.

(ii) Suppose that $X = \Gamma\backslash G/K$ is locally symmetric. Then the same method can be applied to obtain the spectral resolution of the Casimir operator acting on sections of a vector bundle. In this way we recover the results concerning the spectral resolution of the regular representation of G on $L^2(\Gamma\backslash G)$.

The statement concerning the eigenvalues of H which we obtain from Theorem 6.2 can be considerably improved in our case. This is due to the specific geometric structure of the cusp. Let $X = \Gamma\backslash G/K$ be a finite volume locally symmetric space and denote by $N(\lambda)$ the number of eigenvalues of the Casimir operator, belonging to a fixed K-type, which are less than $\lambda \in \mathbb{R}^+$. Borel and Garland [21] proved that $N(\lambda)$ is finite for fixed λ . Recently, Donnelly [28] proved the important result that, for a locally symmmetric space of $\mathbb{Q}$-rank one, $N(\lambda)$ is at most of polynomial growth in λ . This resolves the trace class dilemma for $\mathbb{Q}$-rank one locally symmetric spaces as formulated in [21] and [66]. The method of Donnelly can be easily extended to prove the same result in our case. We give some details of the proof. Donnelly uses modified Neumann comparison.

Let $\varphi \in L^2(X,E)$ and assume that $\varphi = \varphi_1 + \varphi_2$ with $\varphi_1 \in L^2(X_1,E_1)$, $\varphi_2 \in L^2(Y_2,E_2)$. We extend φ_2 by zero to a section of $\mathcal{E}$ over Y. Thus, we may consider φ_2 as an element of $L^2(Y,\mathcal{E})$. Let $(\varphi_2)_o$ be the constant term of φ_2 defined by (2.18). $(\varphi_2)_o$ is contained in $L^2([c,\infty) \times \Gamma_M\backslash X_M, \mathcal{E}_M)$. As above, let $\{\Phi_j\}_{j\in\mathbb{N}}$ be an orthonormal basis of eigenfunctions of Δ_M such that the corresponding sequence of eigenvalues satisfies $\mu_1 \leqq \mu_2 \leqq \cdots$. Let $t \geq 0$. Following Donnelly [28] , we say that φ satisfies the conditions $\mathcal{P}_t$, if for almost all $r \in [c,\infty)$, one has

$$((\varphi_2)_o(r,.),\Phi_j) = 0 \,, \quad \text{for } \mu_j \leqq t - m^2/4 \tag{6.20}$$

Let $\mathcal{H}_t$ be the closed subspace of $L^2(X,E)$ defined by the conditions (6.20). This subspace is invariant under H_o. For $\lambda \geqq 0$ let $N_t(\lambda)$ be the number of eigenvalues of H which are less than λ and correspond to eigenfunctions satisfying the conditions $\mathcal{P}_t$. Further, by $\bar{N}_t(\lambda)$ we shall denote the number of eigenvalues of H_o acting in $\mathcal{H}_t$, which are less than λ . In order to estimate the number of eigenvalues of H we need two auxiliary results. Let $\mu(t)$ be the smallest eigenvalue μ_k of Δ_M such that $m^2/4 + \mu_k > t$. Then, using Theorem 4.38, we get

<u>LEMMA 6.21</u>. The essential spectrum of H_o restricted to $\mathcal{H}_t$ is the half-line $[m^2/4 + \mu(t),\infty)$.

<u>LEMMA 6.22</u>. For $t \geq 0$ and $\lambda \geq 0$ let $M(\lambda,t)$ be the number of eigenvalues of H, which are less than λ and contained in the interval $[t,m^2/4+\mu(t))$. There exists a constant $C > 0$ such that $M(\lambda,t) \leqq C\lambda^{n/2}$ for all $\lambda,t \geq 0$. Here $n = \dim X$.

<u>PROOF</u>. Let $\varphi \in L^2(X,E)$ be an eigenfunction of H with eigenvalue $\omega \in [t,m^2/4+\mu(t))$. Let $\psi \in L^2([c,\infty) \times \Gamma_M\backslash X_M,\mathcal{E}_M)$ be the constant term of $\varphi|Y_2$ defined by (2.18). It follows from (4.25) that ψ satisfies $T\psi = \omega\psi$, T being the operator (4.24). One may expand ψ in a series:

$$\psi(r,x) = \sum_{j=1}^{\infty} a_j(r)\,\Phi_j(x) \quad .$$

Since $T\psi = \omega\psi$, the coefficients a_j satisfy the ordinary differential equation

$$(-r^2 \frac{d^2}{dr^2} + (m-1)r\frac{d}{dr} + \mu_j - \omega)a_j(r) = 0 \quad .$$

If $m^2/4 + \mu_j < \omega$, then the general solution of this equation is

$$C_1 r^{m/2+i\sqrt{\omega-m^2/4-\mu_j}} + C_2 r^{m/2-i\sqrt{\omega-m^2/4-\mu_j}} .$$

Since each a_j is contained in $L^2([c,\infty),r^{-(m+1)}dr)$ it follows that $a_j = 0$ for $m^2/4 + \mu_j < \omega$. This shows that φ satisfies the conditions P_t. We choose $\lambda' < m^2/4+\mu(t)$ as follows: If $\lambda < m^2/4+\mu(t)$ we set $\lambda' = \lambda$. If $\lambda \geq m^2/4+\mu(t)$, then let $\lambda' < m^2/4+\mu(t)$ be greater than the largest eigenvalue of H in $[0,m^2/4+\mu(t))$. Then it follows that $M(\lambda,t) \leq N_t(\lambda')$. According to Lemma 6.21, λ' is smaller than the lower bound of the essential spectrum of $H_o | \mathcal{H}_t$. This allows to employ Neumann comparison to conclude that $N_t(\lambda') \leq \bar{N}_t(\lambda')$ (see [28]). But $\bar{N}_t(\lambda') \leq \bar{N}_t(\lambda)$. Let $\bar{N}(\lambda)$ be the number of eigenvalues of H_o which are less than λ . Since $\mathcal{H}_t$ is a subspace of $L^2(X,E)$, it is clear that $\bar{N}_t(\lambda) \leq \bar{N}(\lambda)$. Moreover, since $H_o = H_1 \oplus H_2$, we have $\bar{N}(\lambda) = \bar{N}_1(\lambda) + \bar{N}_2(\lambda)$, where $\bar{N}_i(\lambda)$ is the number of eigenvalues of H_i, $i=1,2$, which are less than λ . $\bar{N}_1(\lambda)$ can be estimated by standard methods. If we use (2.12), then it is easy to see that the locally invariant operator $\mathcal{D}: C^\infty(Y,E) \to C^\infty(Y,E)$ can be written as

$$\mathcal{D} = r\frac{\partial}{\partial r} \otimes C + \mathcal{D}'_r ,$$

where $\mathcal{D}'_r : C^\infty(\Gamma\backslash Z,E) \longrightarrow C^\infty(\Gamma\backslash Z,E)$ is a first-order operator depending on r. Now consider the space

$$\hat{C}^\infty(X_1,E_1) = \{ \varphi \in C^\infty(X_1,E_1) \mid \nabla_{(\partial/\partial r)}\varphi | \partial X_1 = 0 \} .$$

Let H_ν be the restriction of H to $C^\infty(X_1,E_1)$. We claim that H_1 coincides with Friedrichs' extension of H_ν . Indeed, using the description of $\mathcal{D}$ given above, it follows that, for each $\varphi \in \hat{C}^\infty(X_1,E_1)$, one has $(D\varphi,D\varphi) = (H\varphi,\varphi)$ and, in view of (3.3), we obtain

$$\mathbb{D}[H_\nu] \subset H^1(X_1,E_1) .$$

On the other hand, it is easy to see that $C^\infty(X_1,E_1)$ is a dense subspace of $H^1(X_1,E_1)$. The argument is the same as one uses to prove that $\{f \in \bar{C}^\infty([a,b]) \mid f'(a) = f'(b) = 0\}$ is dense in $H^1([a,b])$. Thus $\mathbb{D}[H_\nu] = H^1(X_1,E_1)$, which proves our claim. Let B be the boundary operator $\nabla_{(\partial/\partial r)}\varphi | \partial X_1$. The pair $(H|X_1,B)$ is a p-elliptic boundary value problem in the sense of [36,p. 184] . Therefore, by [36, Theorem 2.6.1.] , we have

$$\mathrm{Tr}(\exp(-tH_1)) \sim a_1 t^{-n/2} + O(t^{-(n-1)/2})$$

as $t \to 0$. Using a Tauberian theorem [79,p.192], it follows that $\bar{N}_1(\lambda)$ is bounded by a constant multiple of $\lambda^{n/2}$ as $\lambda \to +\infty$. By Proposition 4.39, $\bar{N}_2(\lambda)$ is also bounded by a constant multiple of $\lambda^{n/2}$. Therefore we get $\bar{N}(\lambda) \leqq C\lambda^{n/2}$ for $\lambda > 0$ and a certain constant $C > 0$. Q.E.D.

Now we can state Donnelly's result in our case.

THEOREM 6.23. Let $N(\lambda)$ be the number of eigenvalues of H, acting on $L^2(X,E)$, which are less than λ. Further let $q = \dim X_M$. There exists a constant $C_1 > 0$ such that

$$N(\lambda) \leqq C_1 \lambda^{(n+q)/2}, \quad \lambda > 0 .$$

PROOF. Choose an increasing sequence t_l of real numbers so that the intervals $[t_l, \mu(t_l))$ cover the half-line $[0,\infty)$. In the proof of Proposition 4.39 we have seen that the number of eigenvalues of Δ_M which are less than λ is bounded by a constant multiple of $\lambda^{q/2}$. Thus we may choose a sequence $\{t_l\}_{l\in\mathbb{N}}$ so that the number of intervals $[t_l, \mu(t_l))$ which intersect $[0,\lambda)$ is bounded by a constant multiple of $\lambda^{q/2}$. It follows from Lemma 6.22 that the number of eigenvalues of H in any single interval $[t_l, \mu(t_l))$ is bounded by $C\lambda^{n/2}$. Thus $N(\lambda) \leqq C_1\lambda^{(n+q)/2}$ for $\lambda > 0$ and a certain constant $C_1 > 0$. Q.E.D.

Let $L^2_d(X,E) \subset L^2(X,E)$ be the subspace spanned by the eigenfunctios of H and let $L^2_c(X,E)$ be the absolutely continuous subspace of H. Then we have

$$L^2(X,E) = L^2_d(X,E) \oplus L^2_c(X,E) .$$

Let H_d be the restriction of H to the discrete subspace $L^2_d(X,E)$. From Theorem 6.23 we obtain

COROLLARY 6.24. For each $t > 0$, the heat operator $\exp(-tH_d)$ is of the trace class.

Another consequence of Theorem 6.23 is

COROLLARY 6.25. Let X be a Riemannian manifold with a cusp of rank one. Then the space $\mathcal{H}^*_{(2)}(X)$ of square integrable harmonic forms on X is finite dimensional.

PROOF. Let $\Lambda^*_c(X)$ be the space of C^∞ differential forms on X with compact supports and let $\Delta\colon \Lambda^*_c(X) \longrightarrow \Lambda^*_c(X)$ be the Laplacian. Recall that $\Delta = (d+\delta)^2$ and, as observed at the beginning of
$d+\delta\colon \Lambda^*_c(X) \longrightarrow \Lambda^*_c(X)$ is a generalized Dirac operator. Let $\bar{\Delta}$ be the closure in L^2 of Δ. Then $\mathcal{H}^*_{(2)}(X) = \ker \bar{\Delta}$ and, in view of Theorem 6.23, we have $\dim(\ker \bar{\Delta}) < \infty$. Q.E.D.

COROLLARY 6.26. Let $D : C^\infty(X,E) \longrightarrow C^\infty(X,E)$ be a generalized Dirac operator. Then $\dim(\ker D \cap L^2) < \infty$.

PROOF. Let H be the unique selfadjoint extension of D^2 acting in $L^2(X,E)$ with domain $C^\infty_c(X,E)$. Employing Lemma 5.12, it follows that $\ker H = \ker D \cap L^2$. By Theorem 6.23, we have $\dim(\ker H) < \infty$. Q.E.D.

Now consider a generalized chiral Dirac operator

$$D : C^\infty(X,E) \longrightarrow C^\infty(X,F) .$$

Recall that this means that

$$Q = \begin{pmatrix} 0 & D^* \\ D & 0 \end{pmatrix}$$

is a generalized Dirac operator in the previous sense. In view of Corollary 6.26, $\ker D \cap L^2$ and $\ker D^* \cap L^2$ are both finite dimensional. Therefore, we may define

$$L^2\text{-}\operatorname{Ind} D = \dim(\ker D \cap L^2) - \dim(\ker D^* \cap L^2) \tag{6.27}$$

and we call it the L^2-index of the differental operator D. There are two cases depending on the lower bound of the absolutely continuous spectrum of the Hamiltonian

$$H = \begin{pmatrix} \bar{D}^*\bar{D} & 0 \\ 0 & \bar{D}\bar{D}^* \end{pmatrix}$$

If the lower bound of the absolutely continuous spectrum of H is positive, then the continuous linear extension

$$\bar{D} : H^1(X,E) \to L^2(X,F)$$

of D is a Fredholm operator and the L^2-index of D coincides with the index of $\bar{D}$. On the other hand, if the continuous spectrum of H contains zero, then $\bar{D}$ is not a Fredholm operator. This makes the computation of the L^2-index of D more complicated. However, in most of the applications $\bar{D}$ will be a Fredholm operator.

CHAPTER VII

THE HEAT KERNEL

Let H be as in Ch. VI. In this chapter we shall construct the kernel of the heat operator $\exp(-tH)$, $t > 0$. We employ a variant of the usual parametrix method in the same way as in [62]. An approximate fundamental solution of the heat equation for H can be constructed from the fundamental solution, constructed for the corresponding operator on the cusp in §3, and an interior parametrix.

Following [7], we let $f(a,b)$ denote an increasing C^∞ function of a real variable r such that $f(r) = 0$ for $r \leq a$ and $f(r) = 1$ for $r \geq b$. Define four C^∞ functions Φ_1, Φ_2, Ψ_1, Ψ_2 by

$$\Phi_1 = f(\tfrac{5}{4},\tfrac{3}{2}) \quad , \quad \Psi_1 = f(\tfrac{3}{2},\tfrac{7}{4})$$

$$\Phi_2 = 1 - f(\tfrac{7}{4},2) \quad , \quad \Psi_2 = 1 - \Psi_1 \quad .$$

We regard these functions of r as functions on the cylinder $[1,2] \times N$, where $N = \partial Y_1$ and then extend them to X in the obvious way. Let $\mathcal{E}$ be the locally homogeneous vector bundle over Y which coincides with E over Y_1. Assume that $\mathcal{E}$ is given by the isotropy representation σ of K on V and let $\Delta_\sigma : C_c^\infty(Y,\mathcal{E}) \longrightarrow C_c^\infty(Y,\mathcal{E})$ be the locally invariant operator which coincides with D^2 on Y_1. Let $e_1(z,z',t)$ be the kernel of the heat operator $\exp(-t\Delta_\sigma)$ constructed in Ch. III. As in the proof of Proposition 5.15, the restriction of D^2 to X_1 can be extended to a positive elliptic differential operator $\tilde{H}$ on C^∞ sections of a vector bundle $\tilde{E}$ over a closed Riemannian manifold $\tilde{X}$. Moreover, this construction implies that $\tilde{H}$ is p-elliptic in the sense of [36, p. 167]. Let $\tilde{d}$ be the geodesic distance on $\tilde{X}$. By [36, Th. 1.4.3.], there exists a fundamental solution $e_2(z,z',t)$ for the heat equation $\partial/\partial t + \tilde{H} = 0$ such that, for each $T > 0$, one has

$$\|e_2(z,z',t)\| \leqq C_1 t^{-n/2} \exp(-C_2 \tilde{d}^2(z,z')/t) \qquad (7.1)$$

for some $C_1, C_2 > 0$ and $0 < t \leq T$. If Φ_i, Ψ_i, $i=1,2$, are the functions defined above, we put

$$P(z,z',t) = \sum_{j=1}^{2} \Phi_j(z)e_j(z,z',t)\Psi_j(z') \quad . \tag{7.2}$$

The kernel (7.2) is a parametrix for the fundamental solution of the heat equation for H. Set

$$Q_1(z,z',t) = (\frac{\partial}{\partial t} + H)P(z,z',t) \ , \tag{7.3}$$

where H is applied to the first variable. We introduce the following auxiliary function on X:

$$\tilde{r}(z) = \begin{cases} 1, & \text{if } z \in X - Y_1 \\ r, & \text{if } z \in Y_1 \text{ and } z=(r,x) \ . \end{cases} \tag{7.4}$$

Then we have

(i) $Q_1(z,z',t) = 0$, if $z \notin Y_1 - Y_2$.

(ii) Given $l \in \mathbb{N}$, $T > 0$, there exist constants C_1, C_2 such that

$$\| Q_1(z,z',t)\| \leq C_1 t^l \, \tilde{r}(z)^{m/2} \exp(-C_2 d^2(z,z')/t) \quad ,$$

uniformly for $0 < t \leq T$, $z' \in X$. $d(z,z')$ denotes the geodesic distance of $z,z' \in X$ and m is the number occurring in (2.17).

(i) follows directly from the definition of the kernel P and (ii) is a consequence of Proposition 3.24 and (7.1). Now define Q_k, $k \in \mathbb{N}$, by induction in the usual way by

$$Q_{k+1}(z,z',t) = \int_0^t \int_X Q_1(z,w,t') \circ Q_k(w,z',t-t')\,dw\,dt' \tag{7.5}$$

Observe that $Q_k(z,z',t) \in \mathrm{Hom}(E_{z'}, E_z)$ and the product of the kernels on the right hand side of (7.5) is the composition of homomorphisms. It follows from (i) that $Q_k(z,z',t) = 0$ for $z \notin Y_1 - Y_2$. Hence, the integral (7.5) exists. Using the argument of the proof of Lemma 4.2 in [25] , it follows that the series

$$Q(z,z',t) = \sum_{k=1}^{\infty} (-1)^k Q_k(z,z',t) \tag{7.6}$$

converges uniformly on compacta in the C^∞ topology, and, for each $l \in \mathbb{N}$ and $T > 0$, there exist constants $C_3, C_4 > 0$ such that

$$||Q(z,z',t)|| \leq C_3 t^l \tilde{r}(z')^{m/2} \exp(-C_4 d^2(z,z')/t) \tag{7.7}$$

uniformly for $0 < t \leq T$, $z' \in X$. Moreover $Q(z,z',t) = 0$, if $z \notin Y_1 - Y_2$. Set

$$e(z,z',t) = P(z,z',t) + P{*}Q(z,z',t) \quad , \tag{7.8}$$

where

$$P{*}Q(z,z',t) = \int_o^t \int_X P(z,w,t-t') \circ Q(w,z',t')\,dw\,dt' \quad .$$

Then e is the fundamental solution of the heat equation for the operator H. It follows from (7.7) and the properties of the parametrix that, for each $l \in \mathbb{N}$ and $T > 0$, one has

$$||P{*}Q(z,z',t)|| \leq C_5 t^l (\tilde{r}(z)\tilde{r}(z'))^{m/2} \exp(-C_6 d^2(z,z')/t) \tag{7.9}$$

uniformly for $0 < t \leq T$, $z,z' \in X$. The constants C_5 and C_6 depend on l and T. We summarize the properties of the heat kernel e by

PROPOSITION 7.10. The heat equation for the operator H has a unique fundamental solution $e(z,z',t)$ satisfying the following properties

(i) e is a smooth section of $E \boxtimes E^*$ over $X \times X \times \mathbb{R}^+$ and one has

$$\left(\frac{\partial}{\partial t} + H_z\right) e(z,z',t) = 0 \quad ,$$

H_z being the operator H acting in the first variable.

(ii)

$$\lim_{t \to 0} \int_X e(z,z',t)\varphi(z')\,dz' = \varphi(z) \quad , \qquad \varphi \in C_c^\infty(X,E) \quad .$$

(iii) For each $T > 0$, there exist constants $C_1, C_2 > 0$ such that

$$||e(z,z',t)|| \leq C_1 t^{-n/2} (\tilde{r}(z)\tilde{r}(z'))^{m/2} \exp\left(-C_2 \frac{d^2(z,z')}{t}\right)$$

uniformly for $0 < t \leq T$, z,z' X. Here $\tilde{r}$ is the function (7.4).

The proof of (i) and (ii) is the same as in [17] . (iii) is a consequence of Proposition 3.24, (7.1) and (7.9).

COROLLARY 7.11. The kernel e represents the heat operator $\exp(-tH)$, i.e.

$$(\exp(-tH)\varphi)(z) = \int_X e(z,z',t)\varphi(z')dz'$$

for $\varphi \in L^2(X,E)$.

PROOF. Let $\varphi \in C_c^\infty(X,E)$ and set

$$(A_t\varphi)(z) = \int_X e(z,z',t)\varphi(z')dz' .$$

Using (iii) of Proposition 7.10, it is easy to see that $A_t\varphi$ is contained in $L^2(X,E)$. By (i) and (ii) of this proposition we get $A_t\varphi = \exp(-tH)\varphi$. This shows that A_t extends to a bounded operator on $L^2(X,E)$ and therefore, it coincides with $\exp(-tH)$. Q.E.D.

CHAPTER VIII
THE EISENSTEIN FUNCTIONS

Let H and H_o be the same as in Ch. VI. The purpose of this chapter is to introduce generalized eigenfunctions for the operator H and to express the wave operators $W_{\pm}(H,H_o)$ in terms of the generalized eigenfunctions. In the case of a locally symmetric space a complete system of generalized eigenfunctions is given by the Eisenstein series [56], [66], [74]. A similar system of generalized eigenfunctions for H exists for each manifold with cusps of rank one. We call them **Eisenstein functions**. One of the main features of Eisenstein series is the fact that they can be continued to meromorphic functions in the complex plane. This property is not shared by the Eisenstein functions in general. However, it follows from [38] that there exists a Riemann surface Σ which is a ramified covering of $\mathbb{C}$ of infinite order, so that the Eisenstein functions can be continued to meromorphic functions on Σ.

Let H, H_o and Δ_M be the operators introduced in Ch. VI. They are associated to a generalized Dirac operator $D: C^{\infty}(X,E) \longrightarrow C^{\infty}(X,E)$. Let $\mu_1 < \mu_2 < \cdots$ be the eigenvalues of Δ_M and put

$$\sigma = \{ \mu_j \mid j \in \mathbb{N} \} .$$

For each $\mu \in \sigma$ let $A(\mu)$ denote the corresponding eigenspace of Δ_M. Thus

$$L^2(\Gamma_M \backslash X_M, E_M) = \bigoplus_{\mu \in \sigma} A(\mu) .$$

Put

$$T = \mathbb{C} - [m^2/4 + \mu_1, \infty) . \tag{8.1}$$

For each $\mu \in \sigma$ we denote by $\lambda \longrightarrow \sqrt{\lambda - \mu - m^2/4}$ the branch of the square root whose imaginary part is positive on T. Given $\mu \in \sigma$ and $\lambda \in T$, we define an embedding

$$e_\mu^{\pm}(\lambda)\ :\ A(\mu) \longrightarrow C^\infty(\mathbb{R}^+ \times \Gamma_M\backslash X_M, E_M)$$

by

$$e_\mu^{\pm}(\lambda)(\Phi)(r,x) = r^{m/2 \pm i\sqrt{\lambda-\mu-m^2/4}}\,\Phi(x) \quad , \quad \Phi\in A(\mu). \tag{8.2}$$

For each $\Phi\in A(\mu)$ we have

$$\left(-r^2 \frac{\partial^2}{\partial r^2} + (m-1)r\frac{\partial}{\partial r} + \Delta_M\right)e_\mu^{\pm}(\lambda)(\Phi) = \lambda\, e_\mu^{\pm}(\lambda)(\Phi) \ . \tag{8.3}$$

Via (4.15) we may identify $e_\mu^{\pm}(\lambda)(\Phi)$ with an element of $C^\infty(Y,\mathcal{E})$ and, using (4.25), it follows from (8.3) that

$$(\mathcal{D}^2 - \lambda)e_\mu^{\pm}(\lambda)(\Phi) = 0 \ , \quad \Phi\in A(\mu), \tag{8.4}$$

where $\mathcal{D}$ is the locally invariant operator associated to D. Let $f\in C^\infty(\mathbb{R})$ be such that $f(r) = 0$ for $r \leq 1$ and $f(r) = 1$ for $r \geq 2$. Consider f as a function on Y and put

$$h_\mu(\Phi,\lambda) = (D^2 - \lambda)(fe_\mu^{-}(\lambda)(\Phi)), \tag{8.5}$$

for $\mu\in\sigma$, $\Phi\in A(\mu)$ and $\lambda\in T$. Since D^2 coincides on Y_1 with $\mathcal{D}^2$, (8.4) implies that $h_\mu(\Phi,\lambda)\in C_c^\infty(X,E)$.

<u>DEFINITION 8.6</u>. Let $\mu\in\sigma$ and $\lambda\in T$ .The Eisenstein function associated to $\Phi\in A(\mu)$ with eigenvalue λ is defined as

$$E_\mu(\Phi,\lambda) = fe_\mu^{-}(\lambda)(\Phi) - (H - \lambda)^{-1}(h_\mu(\Phi,\lambda)).$$

<u>LEMMA 8.7</u>. The section $E_\mu(\Phi,\lambda)$ is uniquely determined by the following properties

(i) $E_\mu(\Phi,\lambda)\in C^\infty(X,E)$ and $(D^2 - \lambda)E_\mu(\Phi,\lambda) = 0$ for $\lambda\in T$.

(ii) $E_\mu(\Phi,\lambda) - fe_\mu^{-}(\lambda)(\Phi)\in L^2(X,E)$ for $\Phi\in A(\mu)$, $\lambda\in T$.

Moreover, $E_\mu(\Phi,\lambda)$ is a meromorphic function of $\lambda\in T$ which is holomorphic on $\mathbb{C} - \mathbb{R}^+$.

<u>PROOF</u>. It follows from (8.6) and the elliptic regularity theorem that $E_\mu(\Phi,\lambda)$ satisfies (i) and (ii). The uniqueness is a consequence

of the selfadjointness of H. By Theorem 6.17 and Theorem 4.38, the essential spectrum of H is the half-line $[m^2/4+\mu_1,\infty)$. Therefore, $(H-\lambda)^{-1}$ is a meromorphic function on T (c.f.[72,XII,§4, Lemma 1]). Moreover, as H is a positive operator, its spectrum is contained in $\mathbb{R}^+$. Hence $(H-\lambda)^{-1}$ is holomorphic on $\mathbb{C}-\mathbb{R}^+$. This implies the last statement. Q.E.D.

REMARK 8.8. Let $\Gamma\subset G$ be a lattice of rank one (c.f. example 1 in Ch.V) and let $X=\Gamma\backslash G/K$. Assume that X has a single cusp. Let P be any Γ-percuspidal split parabolic subgroup of G with split component A and Langlands decomposition $P=UAM$. Then $\Gamma\cap P\backslash G/K$ is a cusp of rank one for X. Let E be a locally homogeneous vector bundle over X which is defined by the isotropy representation τ of K on V. Let $\mu\in\sigma$ and $\Phi\in A(\mu)$. Since Φ is an eigenfunction of the Casimir operator, it follows from [21,Corollary 4.3] that Φ is $\mathfrak{Z}(m_{\mathbb{C}})$-finite, and hence, an automorphic form. Extend Φ to a function on $\Gamma\cap P\backslash G$ by $\Phi(uamk)=\tau(k)^{-1}\Phi(m)$, where $u\in U$, $a\in A$, $m\in M$ and $k\in K$. Then, for $s\in\mathbb{C}$ with $\mathrm{Re}(s)>|\rho|$, the Eisenstein series

$$E(P|A:\Phi:s:g)=\sum_{\Gamma\cap P\backslash\Gamma}e^{(s+|\rho|)H(\gamma g)}\Phi(\gamma g)\tag{8.9}$$

converges. We may consider $E(P|A:\Phi:s)$ as an element of $C^\infty(X,E)$. Then it follows from the theory of the constant term [44,II] and Lemma 8.7 that

$$E_\mu(\Phi,s(m-s)+\mu)=E(P|A:\Phi:s)\ .$$

This justifies the name **Eisenstein function** for the section $E_\mu(\Phi,\lambda)$ in the non locally symmetric case.

Basic for the analytic continuation of the Eisenstein series is the theory of the constant term (c.f. [44,IV]). Concerning the constant term of the Eisenstein functions one has the following result

LEMMA 8.10. Let $\mu\in\sigma$. For each $\mu'\in\sigma$ and $\lambda\in T$ there exists a linear operator

$$T_{\mu'\mu}(\lambda):A(\mu)\to A(\mu')$$

which is a meromorphic function of $\lambda\in T$ such that, for each $\Phi\in A(\mu)$, the constant term $E^o_\mu(\Phi,\lambda)$ of $E_\mu(\Phi,\lambda)$ is given by

$$e^-_\mu(\lambda)(\Phi) + \sum_{\mu'\in\sigma} e^+_{\mu'}(\lambda)(T_{\mu'\mu}(\lambda)(\Phi)) .$$

PROOF. The constant term $E^o_\mu(\Phi,\lambda)$ of $E_\mu(\Phi,\lambda)$ can be identified with an element of $C^\infty([1,\infty) \times \Gamma_M\backslash X_M, E_M)$. Using Lemma 8.7 and (4.25), it follows that $E^o_\mu(\Phi,\lambda)$ satifies

$$(-r^2 \frac{\partial^2}{\partial r^2} + (m-1)r\frac{\partial}{\partial r} + \Delta_M)E^o_\mu(\Phi,\lambda) = \lambda E^o_\mu(\Phi,\lambda) \tag{8.11}$$

and $e^-_\mu(\lambda)(\Phi) - E^o_\mu(\Phi,\lambda) \in L^2$. Let $\mu'\in\sigma$ and let $\Phi_1,\ldots,\Phi_p$ be an orthonormal basis for $A(\mu')$. Then it follows from (8.11) that the component of $E^o_\mu(\Phi,\lambda)$ in $L^2([1,\infty))\otimes A(\mu')$ is a linear combination of $e^\pm_{\mu'}(\lambda)(\Phi_i)$, $i=1,\ldots,r$. But only $e^+_{\mu'}(\lambda)(\Phi_i)$ is square integrable on Y_1. Q.E.D.

The operators $T_{\mu'\mu}$ with $\mu'\neq\mu$ are non trivial in general. Therefore, the Eisenstein functions can not be continued analytically to $\mathbb{C}$. However, as observed by L.Guillopé [38] , the Eisenstein functions can be continued to meromorphic functions on the Riemann surface Σ which is associated to the square roots $z\longrightarrow\sqrt{z-\mu-m^2/4}$, $\mu\in\sigma$. Σ is called the spectral surface. It is a ramified covering of $\mathbb{C}$ of infinite order with ramification locus $\sigma_m = \{ m^2/4+\mu \mid \mu\in\sigma \}$. The set T can be identified with one sheet of Σ - the so-called physical sheet. Then, as a consequence of Theorem 7.1 in [38] we obtain the following result

THEOREM 8.12. For each $\mu\in\sigma$ and $\Phi\in A(\mu)$, the Eisenstein function $E_\mu(\Phi,\lambda)$, $\lambda\in T$, has a meromorphic continuation to Σ .

In order to derive Theorem 8.12 from Theorem 7.1 in [38] we have only to verify that the conditions imposed on H and H_o in [38] are satisfied in our case. But this follows from Theorem 4.38. The method employed by L.Guillopé to prove Theorem 7.1 of [38] is an extension of the method used by Colin de Verdiere in the two-dimensional case [22] .

REMARK 8.13. Let $\mu\in\sigma$ and $\Phi\in A(\mu)$. It follows from the construction of the analytic continuation of $E_\mu(\Phi,\lambda)$ in [38] that the section $E_\mu(\Phi,\Lambda) - E^o_\mu(\Phi,\Lambda)$ is square integrable for all $\Lambda\in\Sigma$ different from poles.

In the case of a locally symmetric space it is known that the Eisenstein series satisfy a system of functional equations [44], [56]. The same is true for the Eisenstein functions in general [38]. To describe the result consider the group G of automorphisms of the ramified covering $\Sigma \to \mathbb{C}$. Given $\mu\in\sigma$, let α_μ be a simple closed curve in $\mathbb{C} - \{m^2/4+\mu\}$ which contains no other points of $\sigma_m = \{m^2/4+\mu \mid \mu\in\sigma\}$ in its interior and whose orientation is induced by the standard orientation of $\mathbb{C}$. Denote by $\gamma_\mu \in G$ the element associated to α_μ. Then G is generated by the γ_μ, $\mu\in\sigma$, with the unique relations $\gamma_\mu^2 = 1$ and $\gamma_\mu\gamma_{\mu'}\gamma_\mu^{-1}\gamma_{\mu'}^{-1} = 1$ for $\mu,\mu'\in\sigma$. Then Theorem 8.1 and Theorem 8.2 in [38] imply

<u>THEOREM 8.14</u>. $T_{\mu\mu}(\Lambda)$ is invertible for each $\Lambda\in\Sigma$ and the operators $T_{\mu'\mu}$ satisfy the following functional equations

(i) $T_{\mu\mu}(\gamma_\mu\Lambda) = T_{\mu\mu}^{-1}(\Lambda)$

(ii) $T_{\mu'\mu}(\gamma_\mu\Lambda) = T_{\mu'\mu}(\Lambda)T_{\mu\mu}^{-1}(\Lambda)$, $\mu' \neq \mu$

(iii) $T_{\mu\mu'}(\gamma_\mu\Lambda) = - T_{\mu\mu}^{-1}(\Lambda)T_{\mu\mu'}(\Lambda)$, $\mu' \neq \mu$

(iv) $T_{\nu\mu'}(\gamma_\mu\Lambda) = T_{\nu\mu'}(\Lambda) - T_{\nu\mu}(\Lambda)T_{\mu\mu}^{-1}(\Lambda)T_{\mu\mu'}(\Lambda)$, $\nu,\mu' \neq \mu$,

for $\Lambda\in\Sigma$. Let $\mu,\mu'\in\sigma$, $\mu\neq\mu'$. For each $\Phi\in A(\mu)$ and $\Psi\in A(\mu')$ one has

$$E_\mu(\Phi,\gamma_\mu\Lambda) = E_\mu(T_{\mu\mu}^{-1}(\Lambda)\Phi,\Lambda)$$

$$E_{\mu'}(\Psi,\gamma_\mu\Lambda) = E_{\mu'}(\Psi,\Lambda) - E_\mu(T_{\mu\mu}^{-1}(\Lambda)T_{\mu\mu'}(\Lambda)\Psi,\Lambda)$$

for $\Lambda\in\Sigma$.

Now it is a standard matter to express the wave operators $W_\pm = W_\pm(H,H_0)$ in terms of the Eisenstein functions. Let $f\in C^\infty(\mathbb{R})$ be such that $f(r) = 0$ for $r \leq 2$ and $f(r) = 1$ for $r \geq 3$ and consider f as a function on X in the standard manner. By

$$\tilde{J}\colon L^2(X,E) \longrightarrow L^2(X,E)$$

we shall denote the operator which is the multiplication by f.

<u>LEMMA 8.15</u>. One has

$$\operatorname*{s-lim}_{t\to\pm\infty} (I - \tilde{J})e^{-itH_0}P_0^{ac} = 0 .$$

PROOF. By Theorem 4.38, $P_o^{ac}(L^2(X,E))$ can be identified with the space $L^2([c,\infty)\times\Gamma_M\backslash X_M,E_M)$. It is sufficient to prove

$$\lim_{t\to\pm\infty}(I-\tilde{J})e^{-itH_o}\varphi = 0$$

for $\varphi\in C_c^\infty((c,\infty)\times\Gamma_M\backslash X_M,E_M)$. Let $\hat{\varphi}(\zeta,k)$, $k\in\mathbb{N}$, $\zeta\in\mathbb{R}$, be defined by (4.32). Then we have

$$(e^{-itH_o}\varphi)(r,x) = \tag{8.16}$$

$$= \sum_{k=1}^{\infty}\left(\frac{1}{2\pi}\int_o^\infty \eta(r,\tfrac{m}{2}-i\zeta)e^{-it(m^2/4+\zeta^2+\mu_k)}\hat{\varphi}(\zeta,k)d\zeta\right)\Phi_k(x) ,$$

where $\eta(r,s)$ is the function (4.30) with $b=c$. Observe that $\hat{\varphi}(\zeta,k)$ is rapidly decreasing, i.e., for each $p\in\mathbb{N}$ there exists a constant C_p such that $|\hat{\varphi}(\zeta,k)|\leq C_p(m^2/4+\zeta^2+\mu_k)^{-p}$ for all $k\in\mathbb{N}$, $\zeta\in\mathbb{R}$. Therefore, the right hand side of (8.16) is absolutely convergent. The operator $I-\tilde{J}$ is the multiplication by the function $1-f$, which equals 1 near 2 and 0 on $[3,\infty)$. Hence it is sufficient to estimate (8.16) under the assumption that $2\leq r\leq 3$. For each $\varepsilon>0$ we choose a function $\chi_\varepsilon\in C^\infty(\mathbb{R})$ which satisfies $0\leq\chi_\varepsilon\leq 1$, $\chi_\varepsilon(\zeta)=0$ for $\zeta\leq\varepsilon$ and $\chi_\varepsilon(\zeta)=1$ for $\zeta\geq 2\varepsilon$. Let $\varphi_\varepsilon = J^*(\chi_\varepsilon\hat{\varphi})$, where J is the isometry (4.33). We have

$$\|\varphi-\varphi_\varepsilon\| = \|(1-\chi_\varepsilon)\hat{\varphi}\| \leq 2\varepsilon\|\varphi\| .$$

In (8.16) we replace φ by φ_ε and then estimate the corresponding inegral on the right hand side. Let $v=(\log r)/(2t)$ and assume that $|t|>(\log 3)/\varepsilon$ and $r\in[2,3]$. Integrating by parts we get

$$\left|\int_o^\infty r^{-i\zeta}e^{-it\zeta^2}\chi_\varepsilon(\zeta)\hat{\varphi}(\zeta,k)d\zeta\right| = \left|\int_o^\infty e^{-it(\zeta+v)^2}\chi_\varepsilon(\zeta)\hat{\varphi}(\zeta,k)d\zeta\right| \leq$$

$$\leq |t|^{-1}\left(\int_o^\infty\left(\frac{d}{d\zeta}\left(\frac{\chi_\varepsilon(\zeta)}{\zeta+v}\right)\right)^2 d\zeta\int_o^\infty|\hat{\varphi}(\zeta,k)|^2d\zeta + \frac{1}{\varepsilon^2}\int_o^\infty\left|\frac{d}{d\zeta}\hat{\varphi}(\zeta,k)\right|^2 d\zeta\right) .$$

In the same way one can estimate

$$\int_o^\infty 2^{-i\zeta}r^{i\zeta}e^{-it\zeta^2}\hat{\varphi}(\zeta,k)d\zeta .$$

Since φ has compact support these estimations imply that

$$|(e^{-itH_o}\varphi_\varepsilon)(r,x)| \leq C(\varepsilon)|t|^{-1}\|\varphi\|^2 \quad ,$$

if $r \in [2,3]$ and $|t| > (\log 3)/\varepsilon$. Hence

$$\lim_{t \to \pm\infty} \|(I - \tilde{J})e^{-itH_o}\varphi_\varepsilon\| = 0 \quad .$$

This proves the lemma. Q.E.D.

In view of Lemma 8.15, we may conclude that

$$W_\pm = \text{s-lim}_{t \to \pm\infty} \; e^{itH}\tilde{J}e^{-itH_o}P_o^{ac} \quad .$$

Let

$$W(t) = e^{itH}\tilde{J}e^{-itH_o}P_o^{ac} \quad , \qquad t \geq 0 \; .$$

Assume that $\varphi \in L^2(X,E)$ is such that $P_o{}^{ac}\varphi \in \mathbb{D}(H_o)$. It is easy to see that $\tilde{J}e^{-itH_o}P_o^{ac}\varphi \in \mathbb{D}(H_o)$. Hence $W(t)\varphi$ is a differentiable function of t and

$$\frac{d}{dt}W(t)\varphi = ie^{itH}(H\tilde{J} - \tilde{J}H_o)e^{-itH_o}P_o{}^{ac}\varphi \quad .$$

Integration over a finite interval $[t',t'']$ gives

$$W(t'')\varphi - W(t')\varphi = i\int_{t'}^{t''} (e^{itH}(H\tilde{J} - \tilde{J}H_o)e^{-itH_o}P_o^{ac}\varphi)dt \; .$$

If we pass to the limits $t' \to 0$ and $t'' \to \pm\infty$, then we obtain

$$W_\pm\varphi = \tilde{J}P_o^{ac}\varphi + i\int_0^{\pm\infty}(e^{itH}(H\tilde{J} - \tilde{J}H_o)e^{-itH_o}P_o^{ac}\varphi)dt \quad . \tag{8.17}$$

Now let $\mu \in \sigma$ and $\Phi \in A(\mu)$. For $\lambda \in [m^2/4+\mu,\infty)$ put

$$c_\mu(\lambda) = -\frac{m/2-i\sqrt{\lambda-\mu-m^2/4}}{m/2+i\sqrt{\lambda-\mu-m^2/4}}$$

$$(8.18)$$

$$e_\mu(\lambda)(\Phi) = e_\mu^-(\lambda)(\Phi) + c_\mu(\lambda)e_\mu^+(\lambda)(\Phi)$$

and introduce on the interval $[m^2/4+\mu,\infty)$ the measure

$$d\tau_\mu(\lambda) = \frac{d\lambda}{4\pi\sqrt{\lambda-\mu-m^2/4}} \quad .$$

Given $\hat{g}\in C_o^\infty((m^2/4+\mu,\infty))$, set

$$\varphi = \int_{m^2/4+\mu}^{\infty} e_\mu(\lambda)(\Phi)e^{-it\lambda}\hat{g}(\lambda)d\tau_\mu(\lambda) \quad . \tag{8.19}$$

Since $\hat{g}$ has compact support, $e^{-itH_o}\varphi$ is a C^∞ section. Moreover, the support of $(H\tilde{J} - \tilde{J}H_o)e^{-itH_o}\varphi$ is contained in $[2,3]$. As supp($\hat{g}$) has positive distance from $m^2/4+\mu$, we may proceed as in the proof of Lemma 8.15 and estimate the integral on the right hand side of (8.19). In this way we get

$$\|e^{itH}(H\tilde{J} - \tilde{J}H_o)e^{-itH_o}\varphi\| \leqq C|t|^{-2} \quad , \quad t >> 0 \ .$$

This shows that the integral on the right hand side of (8.17) is absolutely convergent and we may rewrite (8.17) as

$$W_\pm\varphi = \tilde{J}\varphi + i \lim_{\varepsilon\to 0^+} I_\pm(\varepsilon,\varphi) \tag{8.20}$$

where

$$I_\pm(\varepsilon,\varphi) = \int_o^{\pm\infty} e^{\mp\varepsilon t+itH}(H\tilde{J} - \tilde{J}H_o)e^{-itH_o}\varphi\, dt \quad .$$

If we insert (8.19) in $I_\pm(\varepsilon,\varphi)$ and switch the order of integration, then we obtain

$$I_\pm(\varepsilon,\varphi) = \int_{m^2/4+\mu}^{\infty}\hat{g}(\lambda)\{\int_o^{\pm\infty} e^{it(H-\lambda\pm i\varepsilon)}dt\}(D^2-\lambda)(fe_\mu(\lambda)(\Phi))d\tau_\mu(\lambda) =$$

$$= i\int_{m^2/4+\mu}^{\infty} \hat{g}(\lambda)(H-\lambda\pm i\varepsilon)^{-1}(\tilde{h}_\mu(\Phi,\lambda))d\tau_\mu(\lambda) \quad , \tag{8.21}$$

where

$$\tilde{h}_\mu(\Phi,\Lambda) = (D^2 - \Lambda)(fe_\mu(\Lambda)(\Phi)) \quad .$$

Let ∂_+T be the component of the boundary of T which is obtained when $\lambda \in T$ converges to $[m^2/4+\mu_1,\infty)$ through positive imaginary parts. Given $\xi \in T$ and $\eta \in T \cup \partial_+ T$, we introduce the following auxiliary sections

$$E'_\mu(\Phi;\xi,\eta) = fe_\mu(\eta)(\Phi) - (H-\xi)^{-1}(\tilde{h}_\mu(\Phi,\eta)) \quad .$$

Then (8.21) can be rewritten as

$$I_\pm(\varepsilon,\varphi) = i\tilde{J}\varphi - i \int_{m^2/4+\mu}^{\infty} \hat{g}(\lambda)E'_\mu(\Phi;\lambda\mp i\varepsilon,\lambda)d\tau_\mu(\lambda) \quad . \tag{8.22}$$

Using the characterization of the Eisenstein functions given by Lemma 8.7, it follows that $E'_\mu(\Phi;\eta,\eta) = E_\mu(\Phi,\eta)$ for $\eta\in T$ and we get

$$\lim_{\varepsilon\to 0^+} E'_\mu(\Phi;\lambda+i\varepsilon,\lambda) = \lim_{\varepsilon\to 0^+} E'_\mu(\Phi;\lambda+i\varepsilon,\lambda+i\varepsilon) = \lim_{\varepsilon\to 0^+} E_\mu(\Phi,\lambda+i\varepsilon) =$$

$$= E_\mu(\Phi,\lambda) \quad .$$

Now observe that, by Definition 8.6 and Lemma 8.7, one has $\overline{E_\mu(\Phi,\eta)} = E_\mu(\bar{\Phi},\bar{\eta})$ for $\eta\in T$. Using these observations, it follows that

$$\lim_{\varepsilon\to 0^+} E'_\mu(\Phi;\lambda-i\varepsilon,\lambda) = \lim_{\varepsilon\to 0^+} c_\mu(\lambda-i\varepsilon)E'_\mu(\Phi;\lambda-i\varepsilon,\lambda-i\varepsilon) =$$

$$= \lim_{\varepsilon\to 0^+} \overline{c_\mu(\lambda+i\varepsilon)}\,\overline{E_\mu(\Phi,\lambda+i\varepsilon)} = \overline{c_\mu(\lambda)}\,\overline{E_\mu(\bar{\Phi},\lambda)} \quad ,$$

where $c_\mu(\lambda)$ is the function defined by (8.18). If we apply this to (8.22), we get

$$\lim_{\varepsilon\to 0^+} I_-(\varepsilon,\varphi) = i\tilde{J}\varphi - i \int_{m^2/4+\mu}^{\infty} E_\mu(\Phi,\lambda)\hat{g}(\lambda)d\tau_\mu(\lambda)$$

$$\lim_{\varepsilon\to 0^+} I_+(\varepsilon,\varphi) = i\tilde{J}\varphi - i \int_{m^2/4+\mu}^{\infty} \overline{E_\mu(\Phi,\lambda)}c_\mu(\lambda)\hat{g}(\lambda)d\tau_\mu(\lambda) \quad .$$

Combined with (8.20) this leads to

$$W_+\varphi = \int_{m^2/4+\mu}^{\infty} \overline{E_\mu(\Phi,\lambda)}c_\mu(\lambda)\hat{g}(\lambda)d\tau_\mu(\lambda)$$

$$W_-\varphi = \int_{m^2/4+\mu}^{\infty} E_\mu(\Phi,\lambda)\hat{g}(\lambda)d\tau_\mu(\lambda) \tag{8.23}$$

<u>REMARK 8.24</u>. It follows from (8.23) that all Eisenstein functions are regular on $\partial_+ T$.

Now let $\{\Phi_j\}_{j\in\mathbb{N}}$ be an orthonormal basis of eigenfunctions of Δ_M with eigenvalues $\tilde{\mu}_1 \leq \tilde{\mu}_2 \leq \cdots$ and let

$$I: L^2([1,\infty) \times \Gamma_M\backslash X_M, {}_M) \longrightarrow \bigoplus_{j=1}^{\infty} L^2([\tfrac{m^2}{4}+\tilde{\mu}_j,\infty), d\tau_{\tilde{\mu}_j}(\lambda))$$

be the spectral representation of $H_{o,ac}$. Then we have

<u>THEOREM 8.25</u>. Let $(\hat{\varphi}_j) \in \bigoplus_{j=1}^{\infty} C_o^\infty((m^2/4+\tilde{\mu}_j,\infty))$ and $\varphi = I^*((\hat{\varphi}_j))$. Then one has

$$W_+\varphi = \sum_{j=1}^{\infty} \int_{m^2/4+\tilde{\mu}_j}^{\infty} E_{\tilde{\mu}_j}(\Phi_j,\lambda)c_{\tilde{\mu}_j}(\lambda)\hat{\varphi}_j(\lambda)d\tau_{\tilde{\mu}_j}(\lambda)$$

$$W_-\varphi = \sum_{j=1}^{\infty} \int_{m^2/4+\tilde{\mu}_j}^{\infty} E_{\tilde{\mu}_j}(\Phi_j,\lambda)\hat{\varphi}_j(\lambda)d\tau_{\tilde{\mu}_j}(\lambda) \quad .$$

CHAPTER IX
THE SPECTRAL SHIFT FUNCTION

In this chapter we continue with the investigation of the operators H and H_o introduced in Ch. VI. We slightly modify H_o by imposing Neumann boundary conditions at ∂Y_1 in place of ∂Y_2. We shall prove that $\exp(-tH) - \exp(-tH_o)P_o^{ac}$ is of the trace class for each $t > 0$ and our main purpose is to compute the associated spectral shift function which exists by the Krein-Birman theory.

<u>THEOREM 9.1</u>. For each $t > 0$, the operator

$$e^{-tH} - e^{-tH_o}P_o^{ac}$$

is of the trace class. Let $e(z,z',t)$ and $e_o^N(z,z',t)$ denote the kernels of $\exp(-tH)$ and $\exp(-tH_o)$, respectively. Then one has

$$\operatorname{Tr}(e^{-tH} - e^{-tH_o}P_o^{ac}) = \int_X (\operatorname{tr} e(z,z,t) - \operatorname{tr} e_o^N(z,z,t))dz$$

<u>PROOF</u>. Given $t > 0$, set $\tau = t/2$. Then we have

$$e^{-tH} - e^{-tH_o}P_o^{ac} = e^{-\tau H}(e^{-\tau H} - e^{-\tau H_o}P_o^{ac}) + \\ + (e^{-\tau H} - e^{-\tau H_o}P_o^{ac})e^{-\tau H_o}P_o^{ac} \quad . \tag{9.2}$$

Choose $\chi \in C^\infty(X)$ satisfying $0 < \chi \leq 1$, $\chi(z) = 1$ if $z \in X_o$ and $\chi((r,x)) = r^{-1/4}$ if $(r,x) \in Y_2 = [c,\infty) \times \Gamma\backslash Z$, $c = 2^{1/|\lambda|}$. Denote by B_χ the operator of multiplication by χ. The kernel $e(z,z',\tau)$ of $\exp(-\tau H)$ is given by (7.8). The kernel $Q(z,z',t)$ which occurs in (7.8) satisfies $Q(z,z',t) = 0$, if $z \notin Y_1 - Y_2$. Let $d(z)$ denote the distance of z from the boundary ∂Y_2. It follows from (7.7) that there exist constants $C_1, C_2 > 0$ such that

$$\|Q(z,z',t)\| \leq C_1 \tilde{r}(z')^{m/2} \exp(-C_2 \frac{d^2(z')}{t})$$

for $z,z' \in Y_2$ and $0 < t \leq T$. Using Proposition 3.24, it follows that the function $(z,z') \longrightarrow \chi(z)^{-1} \|P*Q(z,z',t)\|$ belongs to $L^2(X\times X)$. Now consider the parametrix P. Let $\Delta: C^\infty(Y,E) \longrightarrow C^\infty(Y,E)$ be the locally invariant operator associated to H and let $e_1(z,z',t)$ be the kernel of $\exp(-t\Delta)$. We may regard the kernel e_1 as a C^∞ function on $\Gamma\backslash G \times \Gamma\backslash G \times \mathbb{R}^+$ with values in End(V) which satisfies the covariance property (3.19). Set

$$\begin{aligned} e_o(g,g',t) &= \int_{\Gamma\cap U\backslash U} e_1(ug,g',t)du \\ \bar{e}_1(g,g',t) &= e_1(g,g',t) - e_o(g,g',t) \quad . \end{aligned} \tag{9.3}$$

The principal symbol of the lifted operator $\tilde{\Delta}$ has the form $\langle\xi,\xi\rangle_x I$. As pointed out by Donnelly in [27, §9], the results of [27] can be extended to operators of this type. In particular, Proposition 5.6 of [27] is valid for $\bar{e}_1(z,z',t)$. Using this proposition and the description (2.16) of the metric on Y, it follows that the function $(z,z') \longrightarrow \chi(z)^{-1} \|\bar{e}_1(z,z',\tau)\|$ is in $L^2(Y_1\times Y_1)$. Next consider the kernel e_o. Since Γ normalizes U, it follows from (9.3) that e_o satisfies $e_o(ug,u'g',t) = e_o(g,g',t)$, $u,u'\in U$. Thus we may identify e_o with a C^∞ section of $E_M \boxtimes E_M^*$ over $(\mathbb{R}^+\times\Gamma_M\backslash X_M)\times(\mathbb{R}^+\times\Gamma_M\backslash X_M)\times\mathbb{R}^+$. Moreover, by (4.25), we get

$$\left(\frac{\partial}{\partial t} - r^2\frac{\partial^2}{\partial r^2} + (m-1)r\frac{\partial}{\partial r} + \Delta_M\right)e_o((r,x),(r',x'),t) = 0 \quad .$$

Thus e_o is the product of the heat kernel for $-r^2\partial^2/\partial r^2 + (m-1)\partial/\partial r$ and the heat kernel $p(x,x',t)$ for Δ_M:

$$\begin{aligned} &e_o((r,x),(r',x'),t) = \\ &= \frac{(rr')^{m/2}}{\sqrt{4\pi t}} \exp\left(-\frac{m^2}{4}t - \frac{(\log(r/r'))^2}{4t}\right)p(x,x',t) \quad . \end{aligned} \tag{9.4}$$

Now consider the kernel $e_o^N(z,z',t)$ of $\exp(-tH_{o,ac})$. According to Corollary 6.18, e_o^N is the continuous part of the heat kernel of the operator

$$-r^2\frac{\partial^2}{\partial r^2} + (m-1)r\frac{\partial}{\partial r} + \Delta_M$$

acting on $C^\infty([1,\infty)\times\Gamma_M\backslash X_M, E_M)$ with Neumann boundary conditions. The discrete part of this heat kernel is given by $m\,p(x,x',t)$. Now it is easy to see that

$$e_o^N((r,x),(r',x'),t) = -\,m\,p(x,x',t) +$$

$$+ \frac{\exp(-\frac{m^2}{4}t)}{\sqrt{4\pi t}} \Big\{ (rr')^{m/2}\exp(-\frac{\log^2(r/r')}{4t}) - \qquad (9.5)$$

$$- \frac{1}{2}\int_{\log(rr')}^{\infty} (m + \frac{u}{t})\exp(\frac{mu}{2} - \frac{u^2}{4t})du \Big\}\, p(x,x',t)$$

Combining our results, it follows that the function

$$(z,z') \longrightarrow \chi(z)^{-1}\, \| e(z,z',\tau) - e_o^N(z,z',\tau)\|$$

belongs to $L^2(X \times X)$. Hence $B_\chi^{-1}(\exp(-\tau H) - \exp(-\tau H_{o,ac})P_o^{ac})$ is a Hilbert-Schmidt operator. Furthermore, (9.5) shows that the function $(z,z') \longrightarrow \|e_o^N(z,z',\tau)\|\chi(z')$ belongs to $L^2(Y_1 \times Y_1)$ and the operator $\exp(-\tau H_{o,ac})P_o^{ac}B_\chi$ is Hilbert-Schmidt. Since

$$e^{-\tau H}B_\chi = (e^{-\tau H} - e^{-\tau H_{o,ac}}P_o^{ac})B_\chi + e^{-\tau H_{o,ac}}P_o^{ac}B_\chi$$

it follows from the observations above that $\exp(-\tau H)B_\chi$ is Hilbert-Schmidt too. Thus the first term on the right hand side of (9.2) is the product of two Hilbert-Schmidt operators, and therefore, a trace class operator. The second term can be treated in the same way. This proves the first statement. The second statement follows from Theorem 4.10 in [77] applied to $X - \partial Y_2$. Q.E.D.

By Corollary 6.24, the operator $\exp(-tH_d) = (I - P^{ac})\exp(-tH)$ is of the trace class for each $t > 0$. Therefore, by Theorem 9.1, it follows that $\exp(-tH)P^{ac} - \exp(-tH_{o,ac})P_o^{ac}$ is of the trace class too and

$$\mathrm{Tr}(e^{-tH} - e^{-tH_o}P_o^{ac}) = \mathrm{Tr}(e^{-tH_d}) + \mathrm{Tr}(e^{-tH}P^{ac} - e^{-tH_o}P_o^{ac}) \qquad (9.6)$$

By the Krein-Birman theory of spectral shift functions [85], the function

$$\xi_t^c(\lambda) = \pi^{-1} \lim_{\varepsilon \to 0^+} \arg\det(I + (e^{-tH}P^{ac} - e^{-tH_o}P_o^{ac})(e^{-tH_o}P_o^{ac} - \lambda - i\varepsilon)^{-1})$$

is a real-valued integrable function with support contained in [0,1] such that

$$\mathrm{Tr}(e^{-tH}P^{ac} - e^{-tH_o}P_o^{ac}) = \int_0^1 \xi_t^c(\lambda)d\lambda$$

(c.f. [85]). By Lemma 3.1 in [85] , the function $-\xi_t^c(e^{-t\lambda})$ in $\lambda > 0$ is independent of $t > 0$. Denote this function by $\xi^c(\lambda;H,H_o)$. This is the spectral shift function associated to H_{ac} , $H_{o,ac}$ and we obtain

$$\mathrm{Tr}(e^{-tH}P^{ac} - e^{-tH_o}P_o^{ac}) = -t\int_0^\infty \xi^c(\lambda;H,H_o)e^{-t\lambda}d\lambda \ . \tag{9.7}$$

Our purpose is now to determine the spectral shift function $\xi^c(\lambda;H,H_o)$. Let $\alpha \in C_o^\infty(\mathbb{R})$. If we proceed along lines similar to the proof of Lemma 3.1 in [85] , it follows that $\alpha(H)P^{ac} - \alpha(H_o)P_o^{ac}$ is of the trace class and

$$\mathrm{Tr}(\alpha(H)P^{ac} - \alpha(H_o)P_o^{ac}) = \int_0^\infty \alpha'(\lambda)\xi^c(\lambda;H,H_o)d\lambda \tag{9.8}$$

and any function in $L^1_{loc}(\mathbb{R})$ with support contained in $[0,\infty)$ which satisfies (9.8) for all $\alpha \in C_o^\infty(\mathbb{R})$ coincides with the spectral shift function ξ^c.

Let $\alpha \in C_o^\infty(\mathbb{R})$. By Theorem 6.17, the wave operator $W_- = W_-(H,H_o)$ satisfies $P^{ac} = W_-W_-^*$ and it intertwines $\alpha(H_{o,ac})$ and $\alpha(H_{ac})$. Thus

$$\alpha(H)P^{ac} = W_-\alpha(H_{o,ac})W_-^* \ . \tag{9.9}$$

Employing Theorem 8.25, it follows from (9.9) that $\alpha(H)P^{ac}$ is an integral operator whose kernel is given by

$$e_\alpha(z,z') = \\ = \sum_{j=1}^\infty \int_{m^2/4+\tilde{\mu}_j}^\infty \alpha(\lambda)E_{\tilde{\mu}_j}(\phi_j,\lambda,z) \otimes \overline{E_{\tilde{\mu}_j}(\phi_j,\lambda,z')}\, d\tau_{\tilde{\mu}_j}(\lambda) \ . \tag{9.10}$$

Observe that this integral-series is actually finite. Thus $e_\alpha(z,z')$ is a smooth kernel. Using the spectral resolution of the operator $H_{o,ac}$ it follows that the kernel of $\alpha(H_{o,ac})$ is given by

$$e_\alpha^o(z,z') =$$

$$= \sum_{j=1}^{\infty} \int_{m^2/4+\tilde{\mu}_j}^{\infty} \alpha(\lambda) e_{\tilde{\mu}_j}(\lambda)(\Phi_j,z) \otimes \overline{e_{\tilde{\mu}_j}(\lambda)(\Phi_j,z')} d\tau_{\tilde{\mu}_j}(\lambda) \quad , \tag{9.11}$$

where $e_{\tilde{\mu}_j}(\lambda)(\Phi_j)$ is defined by (8.18). The kernel $e_\alpha^o(z,z')$ is also smooth and, using Theorem 4.10 of [77] , we obtain

$$\mathrm{Tr}(\alpha(H)P^{ac} - \alpha(H_{o,ac})P_o^{ac}) = \int_X (\mathrm{tr}\, e_\alpha(z,z) - \mathrm{tr}\, e_\alpha^o(z,z))dz \tag{9.12}$$

To compute the right hand side we introduce truncation operators q_b, $b \geq 1$, as follows: Given a locally bounded measurable section φ of E and $b \geq 1$, denote by $(\varphi|Y_b)_o$ the constant term of $\varphi|Y_b$. Then define the truncation operator q_b via the prescription

$$q_b\varphi = \varphi - (\varphi|Y_b)_o \quad . \tag{9.13}$$

q_b satisfies $q_b \circ q_b = q_b$. Assume that $\varphi \in L^2(X,E)$. Then it is clear that $q_b\varphi \in L^2(X,E)$ and $\varphi - q_b\varphi$ is orthogonal to $q_b\varphi$. Thus q_b defines an orthogonal projection of the Hilbert space $L^2(X,E)$. We shall also truncate kernels. Let $h(z,z')$ be a continuous section of $E \boxtimes E^*$ over $X \times X$. For each $z' \in X$, we may regard $z \longrightarrow h(z,z')$ as a section $h_{z'}$ of $E \otimes E^*_{z'}$ and we shall denote by $q_{b,1}h(z,z')$ the section $q_b h_{z'}$. The definition of q_b implies that

$$\text{s-}\lim_{b\to\infty} q_b = \mathrm{Id} \; .$$

Moreover, $q_b(\alpha(H)P^{ac} - \alpha(H_o)P_o^{ac})$ is a trace class operator for each $b \geq 1$ and

$$\lim_{b\to\infty} \mathrm{Tr}(q_b(\alpha(H)P^{ac} - \alpha(H_o)P_o^{ac})) = \mathrm{Tr}(\alpha(H)P^{ac} - \alpha(H_o)P_o^{ac}) \; .$$

On the other hand, $q_{b,1}e_\alpha(z,z')$ and $q_{b,1}e_\alpha^o(z,z')$ are the kernels of $q_b \circ \alpha(H)P^{ac}$ and $q_b \circ \alpha(H_o)P_o^{ac}$, respectively. These kernels are smooth for $z,z' \notin \partial Y_1 \cup \partial Y_b$. Employing again Theorem 4.10 of [77] , we obtain

$$\operatorname{Tr}(\alpha(H)P^{ac} - \alpha(H_o)P_o^{ac}) = \tag{9.14}$$

$$= \lim_{b\to\infty} \int_X (\operatorname{tr} q_{b,1} e_\alpha(z,z) - \operatorname{tr} q_{b,1} e_\alpha^o(z,z))dz$$

Using (9.11), an easy computation gives

$$\int_X \operatorname{tr} q_{b,1} e_\alpha^o(z,z)dz = \sum_{j=1}^{\infty} \int_{m^2/4+\tilde{\mu}_j}^{\infty} \alpha(\lambda)\{ \int_{Y_1-Y_b} \|e_{\tilde{\mu}_j}(\lambda)(\Phi_j,z)\|^2 dz \} d\tau_{\tilde{\mu}_j}(\lambda) =$$

$$= \frac{1}{\pi}\log(b) \sum_{\mu\in\sigma} \dim A(\mu) \int_o^\infty \alpha(\lambda^2 + \frac{m^2}{4} + \mu)d\lambda +$$

$$+ \frac{i}{2\pi} \sum_{\mu\in\sigma} \dim A(\mu) \int_{m^2/4+\mu}^{\infty} \alpha(\lambda) \frac{c'_\mu(\lambda)}{c_\mu(\lambda)} d\lambda +$$

$$+ \frac{1}{2\pi} \sum_{\mu\in\sigma} \dim A(\mu) \int_o^\infty \alpha(\lambda^2 + \frac{m^2}{4} + \mu) \operatorname{Im}(b^{2i\lambda} \cdot \frac{m/2-i\lambda}{m/2+i\lambda}) \frac{d\lambda}{\lambda} .$$

Let $f \in C_o^\infty(\mathbb{R})$. Employing the Riemann-Lebesgue Lemma, it follows that

$$\int_o^\infty f(\lambda) \operatorname{Im}(b^{2i\lambda} \cdot \frac{m/2-i\lambda}{m/2+i\lambda}) \frac{d\lambda}{\lambda} = \frac{\pi}{2} f(0) + o(b)$$

as $b \to \infty$. Thus we obtain

$$\int_X \operatorname{tr} q_{b,1} e_\alpha^o(z,z)dz = \frac{\log(b)}{\pi} \sum_{\mu\in\sigma} \dim A(\mu) \int_o^\infty \alpha(\lambda^2 + \frac{m^2}{4} + \mu)d\lambda +$$

$$+ \frac{i}{2\pi} \sum_{\mu\in\sigma} \dim A(\mu) \int_{m^2/4+\mu}^{\infty} \alpha(\lambda) \frac{c'_\mu(\lambda)}{c_\mu(\lambda)} d\lambda +$$

$$+ \frac{1}{4} \sum_{\mu\in\sigma}' \dim A(\mu)\alpha(\frac{m^2}{4} + \mu) + o(b)$$

as $b \to \infty$. Now we turn to the first integral in (9.14). Set

$$E_{\tilde{\mu}_j}^b(\Phi_j,\lambda) = q_b E_{\tilde{\mu}_j}(\Phi_j,\lambda) , \quad j \in \mathbb{N} .$$

It follows from the construction of the analytic continuation of the Eisenstein functions (c.f. [38]) that the truncated Eisenstein functions are square integrable. Since the integrale-series (9.10) is finite, we obtain

$$\int_X \operatorname{tr} q_{b,1} e_\alpha(z,z)dz = \sum_{j=1}^{\infty} \int_{m^2/4+\tilde{\mu}_j}^{\infty} \alpha(\lambda) \| E^b_{\tilde{\mu}_j}(\Phi_j,\lambda)\|^2 d\tau_{\tilde{\mu}_j}(\lambda) \tag{9.16}$$

Therefore we have to compute the norms of truncated Eisenstein functions. In the locally symmmetric case, the formulas expressing the norms of truncated Eisenstein series by their constant terms are called Maaß-Selberg relations [44]. Employing the same method, one can derive the Maaß-Selberg relations in our case (c.f. [38,§11]). The result is

PROPOSITION 9.17. Let $\mu,\pi \in \sigma$ with $\mu \leqq \pi$ and denote by π^+ the successor of π in σ. Let $\Phi \in A(\mu)$, $b \geq 1$ and set $C_\mu = T_{\mu\mu}$. Then, for each $\tau \in \partial_+ T \cap (m^2/4+\pi, m^2/4+\pi^+)$ one has

$$\begin{aligned}
\| E^b_\mu(\Phi,\lambda)\|^2 &= \log(b)\Big(\sum_{\nu\leqq\pi} \|T_{\nu\mu}(\tau)\Phi\|^2 + \|\Phi\|^2 \Big)\\
&+ i \sum_{\nu\leqq\pi} \sqrt{\tau-\nu-m^2/4}\, \Big\{ \langle T_{\nu\mu}(\tau)\Phi, \frac{\partial T_{\nu\mu}}{\partial\Lambda}(\tau)\Phi\rangle - \langle \frac{\partial T_{\nu\mu}}{\partial\Lambda}(\tau)\Phi, T_{\nu\mu}(\tau)\Phi\rangle \Big\} - \\
&- \frac{1}{2} \sum_{\nu>\pi} \frac{b^{-2\sqrt{\nu+m^2/4-\tau}}}{\sqrt{\nu+m^2/4-\tau}} \|T_{\nu\mu}(\tau)\Phi\|^2 + \operatorname{Im}\left(\frac{b^{2i\sqrt{\tau-\mu-m^2/4}}}{\sqrt{\tau-\mu-m^2/4}} \right) \langle C_\mu(\tau)\Phi,\Phi\rangle
\end{aligned}$$

PROOF. This is a slight modification of the formula given by Corollary 11.3 in [38]. For the convenience of the reader we shall reproduce the proof here. Let $\Phi \in A(\mu)$ and $\lambda \in T$. Using Green's formula, we get

$$\begin{aligned}
(\lambda - \bar{\lambda}) \| E^b_\mu(\Phi,\lambda)\|^2 &= (HE^b_\mu(\Phi,\lambda), E^b_\mu(\Phi,\lambda)) - (E^b_\mu(\Phi,\lambda), HE^b_\mu(\Phi,\lambda)) = \\
&= \langle E^o_\mu(\Phi,\lambda)(b,.), \frac{\partial}{\partial r} E^o_\mu(\Phi,\lambda)(r,.)|_{r=b}\rangle b^{1-m} - \\
&- \langle \frac{\partial}{\partial r} E^o_\mu(\Phi,\lambda)(r,.)|_{r=b}, E^o_\mu(\Phi,\lambda)(b,.)\rangle b^{1-m} .
\end{aligned}$$

We introduce the notation

$$s^\pm_\nu(\lambda) = \sqrt{\lambda-\nu-m^2/4} \pm \sqrt{\bar{\lambda}-\nu-m^2/4}$$

for $\nu \in \sigma$ and $\lambda \in T$. Insert on the right hand side for $E^o_\mu(\Phi,\lambda)$ the expression given by Lemma 8.10 and compute the corresponding scalar products. Then we get

$$(\lambda - \bar{\lambda})\|E^b_\mu(\Phi,\lambda)\|^2 = is^-_\mu(\lambda)b^{-is^+_\mu(\lambda)}\|\Phi\|^2 - i\sum_{\nu\in\sigma} s^-_\nu(\lambda)b^{is^+_\mu(\lambda)}\|T_{\nu\mu}(\lambda)\Phi\|^2 +$$

$$+ is^+_\mu(\lambda)\{b^{-is^-_\mu(\lambda)}(\Phi,C_\mu(\lambda)\Phi) - b^{is^-_\mu(\lambda)}(C_\mu(\lambda)\Phi,\Phi)\}. \tag{9.18}$$

Let $\tau\in\partial_+T \cap (m^2/4+\pi, m^2/4+\pi^+)$ and $\varepsilon > 0$. Observe that, for $\tau > \beta$, one has

$$\sqrt{\tau\pm i\varepsilon-\beta} = \pm\sqrt{\tau-\beta} + i\varepsilon/(2\sqrt{\tau-\beta}) + o(\varepsilon).$$

Hence, for $\tau > \nu+m^2/4$, we obtain

$$\begin{aligned} s^+_\nu(\tau + i\varepsilon) &= i\varepsilon/\sqrt{\tau-\nu-m^2/4} + o(\varepsilon) \\ s^-_\nu(\tau + i\varepsilon) &= 2\sqrt{\tau-\nu-m^2/4} + o(\varepsilon). \end{aligned} \tag{9.19}$$

In the same way we get for $\tau < \nu+m^2/4$

$$\begin{aligned} s^+_\nu(\tau + i\varepsilon) &= 2i\sqrt{\nu+m^2/4-\tau} + o(\varepsilon) \\ s^-_\nu(\tau + i\varepsilon) &= \varepsilon/\sqrt{\nu+m^2/4-\tau} + o(\varepsilon). \end{aligned} \tag{9.20}$$

Now set $\lambda = \tau + i\varepsilon$ in (9.18) and let ε tend to zero. Using (9.19) and (9.20), we obtain

$$\sqrt{\tau-\mu-m^2/4}\,\|\Phi\|^2 = \sum_{\nu\leqq\pi} \sqrt{\tau-\nu-m^2/4}\,\|T_{\nu\mu}(\tau)\|^2. \tag{9.21}$$

Devide (9.18) by $2i\varepsilon$, let ε tend to zero and employ (9.19), (9.20) and (9.21). This gives the desired equation. Q.E.D.

We continue with the investigation of the last term on the right hand side of the equation in Proposition 9.17. According to Remark 8.24 we have

$$\lim_{\tau\to m^2/4+\mu} \|E^b_\mu(\Phi,\tau)\| < \infty.$$

Moreover, $m^2/4 + \mu$ is at most a point of ramification for $C_\mu(\lambda) = T_{\mu\mu}(\lambda)$. Hence

$$\lim_{\tau\to m^2/4+\mu} \left|\sqrt{\tau-\mu-m^2/4}\ \frac{\partial C_\mu}{\partial\Lambda}(\tau)\right| < \infty \quad .$$

Now apply Proposition 9.16 with $\pi=\mu$ and let τ tend to $m^2/4+\mu$. In view of our observations, the last term on the right hand side stays bounded and we get

COROLLARY 9.22. For each $\mu\epsilon\sigma$ one has

$$C_\mu(\frac{m^2}{4}+\mu) = C^*_\mu(\frac{m^2}{4}+\mu) \ .$$

In view of our observations above, it is clear that $\langle C_\mu(\lambda^2+m^2/4+\mu)\Phi,\Phi\rangle$ is a continuous function on $\mathbb{R}$ which is smooth at $\lambda=0$. Let $f\epsilon C_o^\infty(\mathbb{R})$. Using the Riemann-Lebesgue lemma and Corllary 9.22, it follows that

$$\begin{aligned}\int_o^\infty f(\lambda)\,\mathrm{Im}(b^{2i\lambda}\langle C_\mu(\lambda^2+\frac{m^2}{4}+\mu)\Phi,\Phi\rangle)\frac{d\lambda}{\lambda} &= \\ = \frac{\pi}{2}f(0)\langle C_\mu(\frac{m^2}{4}+\mu)\Phi,\Phi\rangle &+ o(b)\end{aligned} \tag{9.23}$$

as $b\longrightarrow\infty$. Let $\mu\epsilon\sigma$ and denote by μ^+ (resp. μ^-) the successor (resp. predecessor) of μ in σ . Let $\alpha\epsilon C_o^\infty(\mathbb{R})$ with support contained in $(m^2/4+\mu^-,m^2/4+\mu^+)$. Then, using (9.16), Proposition 9.17 and (9.23), we obtain

$$\begin{aligned}\int_X \mathrm{tr}\, q_{b,1}e_\alpha(z,z)dz &= \log(b)C(\alpha) + \frac{1}{4}\alpha(\frac{m^2}{4}+\mu)\mathrm{Tr}(C_\mu(\frac{m^2}{4}+\mu)) + \\ &+ \frac{1}{2\pi}\sum_{\nu,\rho\leq\mu}\int_{m^2/4+\mu}^{\infty}\alpha(\tau)\,\mathrm{Im}\{\mathrm{Tr}(T^*_{\nu\rho}(\tau)\frac{\partial T_{\nu\rho}}{\partial\Lambda}(\tau))\}\,d\tau + \\ &+ \frac{1}{2\pi}\sum_{\nu,\rho\leq\mu^-}\int_{-\infty}^{m^2/4+\mu}\alpha(\tau)\,\mathrm{Im}\{\mathrm{Tr}(T^*_{\nu\rho}(\tau)\frac{\partial T_{\nu\rho}}{\partial\Lambda}(\tau))\}\,d\tau + o(b) = \\ &= \log(b)C(\alpha) + \frac{1}{4}\alpha(\frac{m^2}{4}+\mu)\mathrm{Tr}(C_\mu(\frac{m^2}{4}+\mu)) - \\ &- \frac{1}{2\pi}\int_o^\infty \alpha'(\lambda)\left\{\sum_{\nu,\rho\leq\lambda-m^2/4}\int_{m^2/4+\rho}^{\lambda}\mathrm{Im}\{\mathrm{Tr}(T^*_{\nu\rho}(\tau)\frac{\partial T_{\nu\rho}}{\partial\Lambda}(\tau))\}\,d\tau\right\}d\lambda + o(b)\end{aligned} \tag{9.24}$$

as $b \to \infty$. Here $C(\alpha)$ is a constant depending on α. This leads to the main result of this section:

THEOREM 9.25. The spectral shift function $\xi^c(\lambda,H,H_o)$ is given by

$$\xi^c(\lambda;H,H_o) = -\frac{1}{4}\sum_{\mu\leq\lambda-m^2/4} \mathrm{Tr}(C_\mu(\tfrac{m^2}{4}+\mu)) -$$

$$-\frac{1}{2\pi}\sum_{\nu,\mu\leq\lambda-m^2/4}\int_{m^2/4+\mu}^{\lambda} \mathrm{Im}\{\mathrm{Tr}(T^*_{\nu\mu}(\tau)\frac{\partial T_{\nu\mu}}{\partial\Lambda}(\tau))\}\,d\tau +$$

$$+\frac{1}{4}\sum_{\mu\leq\lambda-m^2/4}\dim A(\mu) +$$

$$+\frac{1}{\pi m}\sum_{\mu\leq\lambda-m^2/4}\dim A(\mu)\arctan(\tfrac{2}{m}\sqrt{\lambda-m^2/4-\mu}) .$$

PROOF. Denote by $\tilde{\xi}^c(\lambda)$ the right hand side of the equation claimed in the theorem. Let $\rho\in\sigma$ and $\alpha\in C_o^\infty(\mathbb{R})$ with $\mathrm{supp}\,\alpha\subset(\rho-\varepsilon,\rho+\varepsilon)$ so that $\sigma-\{\rho\}\cap(\rho-\varepsilon,\rho+\varepsilon) = \emptyset$. Using (9.15) and (9.24), an easy computation shows that

$$\int_X \{\mathrm{tr}\, q_{b,1}e_\alpha(z,z) - \mathrm{tr}\, q_{b,1}e^o_\alpha(z,z)\}dz = \log(b)C_1(\alpha) +$$

$$+\int_o^\infty \alpha'(\lambda)\,\tilde{\xi}^c(\lambda)d\lambda + o(b)$$

as $b \to \infty$. Here $C_1(\alpha)$ is a constant depending on α. In view of (9.14), the left hand side of this equation stays bounded as $b \to \infty$. Hence $C_1(\alpha) = 0$ and, by (9.14), we obtain

$$\mathrm{Tr}(\alpha(H)P^{ac} - \alpha(H_o)P_o^{ac}) = \int_o^\infty \alpha'(\lambda)\tilde{\xi}^c(\lambda)d\lambda .$$

This equation extends obviously to all $\alpha\in C_o^\infty(\mathbb{R})$. Now observe that $\tilde{\xi}^c$ is locally integrable with support contained in $[0,\infty)$. Hence (9.8) implies that $\tilde{\xi}^c(\lambda) = \xi^c(\lambda;H,H_o)$. Q.E.D.

REMARK 9.26. Theorem 9.25 shows that the spectral shift function $\xi^c(\lambda;H,H_o)$ is piecewise continuous.

CHAPTER X
THE L^2-INDEX OF GENERALIZED DIRAC OPERATORS

We are now ready to derive the first version of an index formula for generalized chiral Dirac operators. Let

$$D: C^\infty(X,E^+) \longrightarrow C^\infty(X,E^-)$$

be a generalized chiral Dirac operator in the sense of Definition 5.3'. Consider the operator

$$\tilde{D} = \begin{pmatrix} 0 & D^* \\ D & 0 \end{pmatrix} : \begin{matrix} C^\infty(X,E^+) \\ \oplus \\ C^\infty(X,E^-) \end{matrix} \longrightarrow \begin{matrix} C^\infty(X,E^+) \\ \oplus \\ C^\infty(X,E^-) \end{matrix}$$

Set $E = E^+ \oplus E^-$. Then we may regard $\tilde{D}$ as a generalized Dirac operator on $C^\infty(X,E)$. Let Q be the closure of $\tilde{D}$ in L^2. Q is a selfadjoint operator in the Hilbert space $L^2(X,E)$ and the Hamiltonian considered in the previous section is given by $H = Q^2$. Written in matrix form we have

$$H = \begin{pmatrix} H^+ & 0 \\ 0 & H^- \end{pmatrix}$$

where H^+ (resp. H^-) is the unique selfadjoint extension of D^*D (resp. DD^*). Thus the results of the previous sections apply to H^+ and H^-, respectively. The L^2-index of D defined by (6.27) is finite. Moreover, employing Corollary 6.24 and adapting the argument used on p. 297 in [5], we obtain

$$L^2\text{-Ind}(D) = \mathrm{Tr}(e^{-tH_d^+}) - \mathrm{Tr}(e^{-tH_d^-}) \qquad (10.1)$$

Now we may continue by employing the results of Ch. IX. However, the scattering system (H,H_0) considered in Ch. IX is not supersymmetric (c.f. Ch. I) which is important for our purpose. Therefore, we shall modify the free Hamiltonian H_0.

Let $\mathcal{E}^\pm$ be the locally homogeneous vector bundles over Y associated to $E^\pm$ and denote by $\sigma^\pm: K \to GL(V^\pm)$ the isotropy representations of $\tilde{E}^{\pm}$. Furthermore, let

$$\mathcal{D}: C^{\infty}(Y, E^{+}) \longrightarrow C^{\infty}(Y, E^{-})$$

be the locally invariant operator associated to D. By (5.4), there exist $L^{\pm}$ $\mathrm{End}_K(V^{\pm})$ such that

$$\tilde{\mathcal{D}}^* \tilde{\mathcal{D}} = - R(\Omega) \otimes \mathrm{Id}_{V^+} + \mathrm{Id} \otimes L^+$$

$$\tilde{\mathcal{D}}\tilde{\mathcal{D}}^* = - R(\Omega) \otimes \mathrm{Id}_{V^-} + \mathrm{Id} \otimes L^- .$$

Let Δ^+ (resp. Δ^-) be the unique selfadjoint extension of $\mathcal{D}^*\mathcal{D}$ (resp. $\mathcal{D}\mathcal{D}^*$) acting on $C_c^{\infty}(Y, E^{\pm})$. Further, let $E_M^{\pm}$ be the locally homogeneous vector bundles over $\Gamma_M \backslash X_M$ obtained by restricting the K-modules $V^{\pm}$ to K_M. Recall that the constant term φ_o of any section $\varphi \in C^{\infty}(Y, E^{\pm})$ is contained in the space $(C^{\infty}((\Gamma U)\backslash G) \otimes V^{\pm})^K \simeq C^{\infty}(\mathbb{R}^+ \times \Gamma_M \backslash X_M, E_M^{\pm})$. We observe that $\mathcal{D}$ maps $(C^{\infty}((\Gamma U)\backslash G) \otimes V^+)^K$ to $(C^{\infty}((\Gamma U)\backslash G) \otimes V^-)^K$. Denote by $\mathcal{D}_o$ the induced operator

$$\mathcal{D}_o: C^{\infty}(\mathbb{R}^+ \times \Gamma_M \backslash X_M, E_M^+) \longrightarrow C^{\infty}(\mathbb{R}^+ \times \Gamma_M \backslash X_M, E_M^-) .$$

Let $(\mathcal{D}^*)_o$ be the restriction of $\mathcal{D}^*$ to $(C^{\infty}((\Gamma U)\backslash G) \otimes V^-)^K$. Then we have $(\mathcal{D}^*)_o = (\mathcal{D}_o)^*$ and we shall simply write $\mathcal{D}_o^*$ for this operator. Recall from Ch. IV that there exist elliptic operators $\Delta_M^{\pm}$ acting on $C^{\infty}(\Gamma_M \backslash X_M, E_M^{\pm})$ such that

$$\mathcal{D}_o^* \mathcal{D}_o = - r^2 \frac{\partial^2}{\partial r^2} + (m-1) r \frac{\partial}{\partial r} + \Delta_M^+$$

$$\mathcal{D}_o \mathcal{D}_o^* = - r^2 \frac{\partial^2}{\partial r^2} + (m-1) r \frac{\partial}{\partial r} + \Delta_M^- \qquad (10.2)$$

<u>LEMMA 10.3</u>. There exists a bundle isomorphism $\eta: E_M^+ \rightarrow E_M^-$ and a selfadjoint operator

$$D_M: C^{\infty}(\Gamma_M \backslash X_M, E_M^+) \longrightarrow C^{\infty}(\Gamma_M \backslash X_M, E_M^-)$$

such that

(i) $\mathcal{D}_o = \eta(r\frac{\partial}{\partial r} + D_M)$, $\mathcal{D}_o^* = (mI - r\frac{\partial}{\partial r} + D_M)\eta^{-1}$.

(ii) $\Delta_M^+ = D_M{}^2 + mD_M$, $\Delta_M^- = \eta \cdot \Delta_M^+ \cdot \eta^{-1}$.

PROOF. It is clear that $\mathcal{D}_o$ is a locally invariant operator. Since $\mathcal{D}_o$ is a first-order operator, there exist $B \in \mathrm{Hom}(V^+, V^-)$ and a locally invariant operator

$$L_M: C^\infty(\Gamma_M \backslash X_M, E_M^+) \longrightarrow C^\infty(\Gamma_M \backslash X_M, E_M^-)$$

such that $\mathcal{D}_o = r\frac{\partial}{\partial r} \otimes B + L_M$ (c.f.[60]) . To compute $\mathcal{D}_o^*$ we simply observe that the volume element of $\mathbb{R}^+ \times \Gamma_M \backslash X_M$ is given by $dvol = = r^{-(m+1)} dr\, dvol(x)$. This gives

$$\mathcal{D}_o^* = (mI - r\frac{\partial}{\partial r}) \otimes B^* + L_M^* \ .$$

Hence, we obtain

$$\mathcal{D}_o^* \mathcal{D}_o = (-r^2 \frac{\partial^2}{\partial r^2} + (m-1) r \frac{\partial}{\partial r}) \otimes B^* B + L_M^* L_M + m B^* L_M +$$

$$+ r\frac{\partial}{\partial r} \otimes (L_M^* \circ B - B^* \circ L_M) \ .$$

Combined with (10.2) this leads to

$$B^* B = B B^* = Id \ , \quad \Delta_M^+ = L_M^* L_M + m B^* L_M \quad \text{and} \quad L_M^* B = B^* L_M \ .$$

Set $D_M = B^* L_M$ and let $\eta: E_M^+ \longrightarrow E_M^-$ be the bundle map induced by B. Since $BB^* = Id$, η is an isomorphism and $\mathcal{D}_o = \eta(r\frac{\partial}{\partial r} + D_M)$. The Lemma follows from the relations above. Q.E.D.

Define a map $F: C^\infty(\mathbb{R}^+) \to C^\infty(\mathbb{R})$ by $F(f) = e^{-(m/2)u} f(e^u)$. If $f \in L^2(\mathbb{R}^+, r^{-(m+1)} dr)$, then $F(f) \in L^2(\mathbb{R}, du)$ and F defines an isometry

$$F: L^2(\mathbb{R}^+, r^{-(m+1)} dr) \longrightarrow L^2(\mathbb{R}, du) \ . \tag{10.4}$$

For $f \in C^\infty(\mathbb{R}^+)$ one has

$$F(r\frac{d}{dr} f) = (\frac{d}{du} + \frac{m}{2} I) F(f) \ .$$

Put

$$\tilde{D}_M = D_M + \frac{m}{2} I \ .$$

By Lemma 10.3 we have $\mathcal{D}_o = \eta F^{-1}(\frac{\partial}{\partial u} + \tilde{D}_M) F$, $\mathcal{D}_o^* = F^{-1}(-\frac{\partial}{\partial u} + \tilde{D}_M) F \eta^{-1}$

and $\tilde{D}_M$ is a selfadjoint operator on $C^\infty(\Gamma_M\backslash X_M, E_M^+)$. Let P denote the projection of $C^\infty(\Gamma_M\backslash X_M, E_M^+)$ onto the subspace spannned by all eigenfunctions Φ_ω of $\tilde{D}_M$ with eigenvalues $\omega \geq 0$. Consider the operator

$$\frac{\partial}{\partial u} + \tilde{D}_M \tag{10.5}$$

acting on $C^\infty(\mathbb{R}^+ \times \Gamma_M\backslash X_M, E_M^+)$. Following [7] , we introduce the boundary conditions

$$Pf(0,.) = 0 \quad . \tag{10.6}$$

The formal adjoint operator of (10.5) is

$$-\frac{\partial}{\partial u} + \tilde{D}_M \tag{10.7}$$

and the adjoint boundary conditions to (10.6) are

$$(I - P)f(0,.) \quad . \tag{10.8}$$

Let Q_o^+ (resp. Q_o^-) be the closure in L^2 of the operator defined by $\partial/\partial u + \tilde{D}_M$ (resp. $-\partial/\partial u + \tilde{D}_M$) with domains given by (10.6) (resp. (10.8)). By Proposition 2.12 of [7] , Q_o^+ and Q_o^- are adjoint of each other. Set

$$Q_o = \begin{pmatrix} 0 & Q_o^- \\ Q_o^+ & 0 \end{pmatrix}$$

Then Q_o is a selfadjoint operator in $L^2(\mathbb{R}^+ \times \Gamma_M\backslash X_M, E_M^+ \oplus E_M^+)$. Set

$$\tilde{H}_o = Q_o^2 \quad .$$

Using the isometry $\begin{pmatrix} 0 & F\eta^{-1} \\ F & 0 \end{pmatrix}$ we may regard $\tilde{H}_o$ as a selfadjoint operator in $L^2([1,\infty) \times \Gamma_M\backslash X_M, E_M^+ \oplus E_M^-)$. This is our modified free Hamiltonian. Observe that

$$\tilde{H}_o = \begin{pmatrix} \tilde{H}_o^+ & 0 \\ 0 & \tilde{H}_o^- \end{pmatrix}$$

where $\tilde{H}_o^\pm = Q_o^\mp Q_o^\pm$.

Next we shall compare $\tilde{H}_o$ with the Hamiltonian H_o used in Ch. IX. Let H_o^+ be the selfadjoint operator in $L^2(\mathbb{R}^+ \times \Gamma_M\backslash X_M, E_M^+)$ defined by

$-\partial^2/\partial u^2 + \Delta_M^+ + (m^2/4)I$ with domain the space of compactly supported C^∞ sections satisfying the boundary conditions $(\frac{\partial}{\partial u} + \frac{m}{2})f|_{u=0} = 0$. Then via the isometry $\begin{pmatrix} 0 & F\eta^{-1} \\ F & 0 \end{pmatrix}$, $H_{o,ac}$ is unitarily equivalent to

$$\begin{pmatrix} H^+_{o,ac} & 0 \\ 0 & H^+_{o,ac} \end{pmatrix}$$

LEMMA 10.9. For each $t > 0$, the operator

$$\exp(-tH^+_{o,ac}) - \exp(-t\tilde{H}^{\pm}_o)$$

is of the trace class.

PROOF. Let $p^+(x,x',t)$ denote the kernel of $\exp(-t\Delta_M^+)$. Similar to (9.5) we obtain the following expression for the kernel of $\exp(-tH^+_{o,ac})$

$$\frac{\exp(-\frac{m^2}{4}t)}{\sqrt{4\pi t}}\{\exp(-\frac{(u-v)^2}{4t}) -$$

$$-\frac{1}{2}\exp(-\frac{m}{2}(u+v))\int_{u+v}^{\infty}(m+\frac{w}{t})\exp(\frac{mw}{2}-\frac{w^2}{4t})dw\}\, p^+(x,x',t) -$$

$$-\, m\exp(-\frac{m}{2}(u+v)\, p^+(x,x',t) \quad .$$

Further, observe that $\tilde{H}^+_o$ is the operator given by $-\partial^2/\partial u^2 + \tilde{D}^2_M$ with the boundary conditions

$$Pf(0,.) = 0 \quad \text{and} \quad (I-P)\{(\frac{\partial f}{\partial u} + \tilde{D}_M f)_{u=0}\} = 0 \ . \qquad (10.10)$$

and $\tilde{H}^-_o$ is obtained in the same manner with the boundary conditions

$$P\{(-\frac{\partial f}{\partial u} + \tilde{D}_M f)_{u=0}\} = 0 \quad \text{and} \quad (I-P)f(0,.) = 0 \ . \qquad (10.11)$$

The heat kernels of these operators are described on p. 52 in [7]. Finally observe that by Lemma 10.3, one has $\tilde{D}_M{}^2 = \Delta_M^+ + (m^2/4)I$. Now we can proceed along lines similar to the proof of Theorem 9.1 and the lemma follows. Q.E.D.

This lemma combined with Theorem 9.1 shows that the operator

$$\exp(-tH) - \exp(-t\tilde{H}_o)P_o^{ac} \qquad (10.12)$$

is of the trace class. There exists a spectral shift function $\xi^c(\lambda;H,\tilde{H}_o)$ so that formula (9.7) is valid with H_o replaced by $\tilde{H}_o$. Moreover, the restriction of the operator (10.12) to $L^2(X,E^\pm)$ is is also of trace class. Let $P_\pm$ denote the orthogonal projection of $L^2(X,E)$ onto $L^2(X,E^\pm)$. Then, similar to (9.7), there exist spectral shift functions $\xi_+^c(\lambda;H,\tilde{H}_o)$ and $\xi_-^c(\lambda;H,\tilde{H}_o)$ so that

$$\xi^c(\lambda;H,\tilde{H}_o) = \xi_+^c(\lambda;H,\tilde{H}_o) + \xi_-^c(\lambda;H,\tilde{H}_o)$$

and

$$\mathrm{Tr}(P_\pm(\exp(-tH)P^{ac} - \exp(-t\tilde{H}_o)P_o^{ac})) = -t\int_0^\infty \xi_\pm^c(\lambda;H,\tilde{H}_o)e^{-t\lambda}d\lambda \quad . \tag{10.13}$$

Put

$$\tau = P_+ - P_- \, . \tag{10.14}$$

Employing (9.6) and (10.13), formula (10.1) can be rewritten as

$$\begin{aligned} L^2\text{-Ind}(D) &= \mathrm{Tr}(\tau(\exp(-tH) - \exp(-t\tilde{H}_o)P_o^{ac})) \; + \\ &\quad + t\int_0^\infty (\xi_+^c(\lambda;H,\tilde{H}_o) - \xi_-^c(\lambda;H,\tilde{H}_o))e^{-t\lambda}d\lambda \; . \end{aligned} \tag{10.15}$$

We continue with the investigation of the function $\xi_+^c(\lambda;H,\tilde{H}_o)$ - - $\xi_-^c(\lambda;H,\tilde{H}_o)$. By Lemma 10.3,(ii), the eigenvalues of Δ_M^+ and Δ_M^- coincide. We denote this set by σ. Given $\mu\in\sigma$, let $A^\pm(\mu)$ be the eigenspace of $\Delta_M^\pm$ with eigenvalue μ. The isomorphism $\eta: E_M^+ \xrightarrow{\simeq} E_M^-$ which exists by Lemma 10.3 induces an isomorphism $\eta: A^+(\mu) \xrightarrow{\simeq} A^-(\mu)$. According to §8, to each $\Phi\in A^\pm(\mu)$ there is associated an Eisenstein function $E_\mu^\pm(\Phi,\lambda)$ which is a generalized eigenfunction of $H^\pm$. Denote by $T_{\mu',\mu}^\pm(\lambda): A^\pm(\mu) \longrightarrow A^\pm(\mu')$, $\mu'\in\sigma$, the operators associated to $E_\mu^\pm(\lambda)$ via Lemma 8.10 and set $C_\mu^\pm = T_{\mu\mu}^\pm$. Let $\xi_\pm^c(\lambda;H,\tilde{H}_o)$ be the function defined by the formula of Theorem 9.25 with C_μ, $T_{\nu\mu}$ and $A(\mu)$ replaced by $C_\mu^\pm$, $T_{\nu\mu}^\pm$ and $A^\pm(\mu)$, respectively. Then we have

$$\xi_\pm^c(\lambda;H,\tilde{H}_o) = \xi_\pm^c(\lambda;H,H_o) + \xi_\pm^c(\lambda;H_o,\tilde{H}_o) \tag{10.16}$$

where $\xi_+^c(\lambda;H_o,\tilde{H}_o)$ and $\xi_-^c(\lambda;H_o,\tilde{H}_o)$ are the spectral shift functions associated to $(H_{o,ac}^+,\tilde{H}_o^+)$ and $(H_{o,ac}^+,\tilde{H}_o^-)$, respectively. To compute $\xi_\pm^c(\lambda;H_o,\tilde{H}_o)$ we may proceed along lines similar to Ch. IX. Given $\alpha\in C_o^\infty(\mathbb{R})$

denote by $\tilde{e}_\alpha^{\pm}(z,z')$ the kernel of the integral operator $\alpha(\tilde{H}_o^{\pm})$. The kernel $\tilde{e}_\alpha^{\pm}(z,z')$ has a description in terms of generalized eigenfunctions similar to (9.11). To describe the generalized eigenfunctions of $\tilde{H}_o^{\pm}$ observe that $\tilde{D}_M$ has a discrete spectrum with real eigenvalues ω and eigenfunctions Φ_ω. Given an eigenvalue ω of $\tilde{D}_M$, set

$$c_\omega^{\pm}(\lambda) = \begin{cases} -1 & , \quad \pm\omega \geqq 0 \\ -\dfrac{\omega \mp i\sqrt{\lambda-m^2/4-\omega^2}}{\omega \pm i\sqrt{\lambda-m^2/4-\omega^2}} & , \quad \pm\omega < 0 \end{cases}$$

Then, by (1o.10) and (10.11), the generalized eigenfunctions of $\tilde{H}_o^{\pm}$ are given by

$$\left(e^{i\sqrt{\lambda-m^2/4-\omega^2}\,u} + c_\omega^{\pm}(\lambda)e^{-i\sqrt{\lambda-m^2/4-\omega^2}\,u}\right)\Phi_\omega \qquad ,$$

where ω runs over the eigenvalues of $\tilde{D}_M$. Next we have to compute the integral

$$\int_o^b \int_{\Gamma_M\backslash X_M} \operatorname{tr} \tilde{e}_\alpha^{\pm}((r,x),(r,x))drdx \quad .$$

Finally, observe that

$$\int_o^\infty \alpha'(\lambda)\left\{\xi_+^c(\lambda;H_o,\tilde{H}_o) - \xi_-^c(\lambda;H_o,\tilde{H}_o)\right\}d\lambda =$$

$$= \lim_{b\to\infty} \int_o^b \int_{\Gamma_M\backslash X_M} \left\{\operatorname{tr} \tilde{e}_\alpha^-((r,x),(r,x)) - \operatorname{tr} \tilde{e}_\alpha^+((r,x),(r,x))\right\}drdx \ .$$

If we carry out these computations we obtain

$$\xi_+^c(\lambda;H_o,\tilde{H}_o) - \xi_-^c(\lambda;H_o,\tilde{H}_o) = -\frac{1}{2}\dim(\ker\tilde{D}_M) + \tag{10.17}$$
$$+ \frac{1}{\pi} \sum_{\substack{\omega^2 \leqq \lambda-m^2/4 \\ \omega\neq 0}} m(\omega)\operatorname{sign}\omega \arctan\left(\frac{1}{|\omega|}\sqrt{\lambda-m^2/4-\omega^2}\right)$$

where $m(\omega)$ denotes the multiplicity of the eigenvalue ω. Employing (10.16), (10.17) and using the definition of $\xi_\pm^c(\lambda;H,H_o)$ we obtain

$$\xi_+^c(\lambda;H,\tilde{H}_o) - \xi_-^c(\lambda;H,\tilde{H}_o) = \frac{1}{2}\dim(\ker\tilde{D}_M) - \tag{10.18}$$
$$- \frac{1}{4}\sum_{\mu\leq\lambda-m^2/4}\{\mathrm{Tr}(C_\mu^+(\frac{m^2}{4}+\mu)) - \mathrm{Tr}(C_\mu^-(\frac{m^2}{4}+\mu))\} + F(\lambda)$$

where $F(\lambda)$ is a continuous function on $\mathbb{R}^+$ with $F(\lambda)=0$ for $\lambda\notin\sigma_{ac}(H)$. In order to simplify this expression further we need the following auxiliary result

LEMMA 10.19. For each $\mu\in\sigma$ one has

$$\eta(\tilde{D}_M + i\sqrt{\lambda-\mu-m^2/4}\,I)C_\mu^+(\lambda) = C_\mu^-(\lambda)\eta(\tilde{D}_M - i\sqrt{\lambda-\mu-m^2/4}\,I)$$

$$(\tilde{D}_M - i\sqrt{\lambda-\mu-m^2/4}\,I)\,\eta^{-1}C_\mu^-(\lambda) = C_\mu^+(\lambda)(\tilde{D}_M + i\sqrt{\lambda-\mu-m^2/4}\,I)\,\eta^{-1}.$$

PROOF. Let $\Phi\in A^+(\mu)$, $\lambda\in T$ and consider the section $DE_\mu^+(\Phi,\lambda)$. By Lemma 8.7, we have $(DD^*-\lambda)DE_\mu^+(\Phi,\lambda)=0$. Using Definition 8.6 and Lemma 5.12 it follows that $DE_\mu^+(\Phi,\lambda) - D(fe_\mu^-(\lambda)(\Phi))$ belongs to $L^2(X,E^-)$. By Lemma 8.10 and Lemma 10.3, the constant term of $DE_\mu^+(\Phi,\lambda)$ is given by

$$e_\mu^-(\lambda)(\eta(D_M+(\frac{m}{2}-i\sqrt{\lambda-\mu-m^2/4}))\Phi) +$$
$$\sum_{\mu'\in\sigma} e_\mu^+(\lambda)(\eta(D_M + (\frac{m}{2}+i\sqrt{\lambda-\mu'-m^2/4}))T_{\mu',\mu}^+(\lambda)\Phi) \quad .$$

Employing Lemma 8.7, we obtain

$$DE_\mu^+(\Phi,\lambda) = E_\mu^-(\eta(D_M + (\frac{m}{2}-i\sqrt{\lambda-\mu-m^2/4}))\Phi,\lambda) \quad .$$

Comparing the constant terms of the left and the right hand side, we get the first equation. Similar considerations for D^* lead to the second equation. Q.E.D.

COROLLARY 10.20. Let $\mu\in\sigma$. Then one has

(i) $\mathrm{Tr}(C_\mu^+(\frac{m^2}{4}+\mu)) = \mathrm{Tr}(C_\mu^-(\frac{m^2}{4}+\mu))$, if $\frac{m^2}{4}+\mu\neq 0$

(ii) $\mathrm{Tr}(C_\mu^+(\frac{m^2}{4}+\mu)) = -\mathrm{Tr}(C_\mu^-(\frac{m^2}{4}+\mu))$, if $\frac{m^2}{4}+\mu = 0$.

PROOF. Let $\lambda\in T$. By Lemma 10.3, we have

$$(\tilde{D}_M + i\sqrt{\lambda-\mu-m^2/4}\, I)(\tilde{D}_M - i\sqrt{\lambda-\mu-m^2/4}\, I) = \Delta_M^+ + (\lambda-\mu)I \ .$$

Employing Lemma 10.3 and Lemma 10.19, it follows that

$$\lambda \mathrm{Tr}(C_\mu^-(\lambda)) = \mathrm{Tr}((\Delta_M^- + (\lambda-\mu)I)C_\mu^-(\lambda)) =$$

$$= \mathrm{Tr}(\eta(\tilde{D}_M + i\sqrt{\lambda-\mu-m^2/4}\, I)C_\mu^+(\lambda)(\tilde{D}_M + i\sqrt{\lambda-\mu-m^2/4}\, I)\eta^{-1}) =$$

$$= \lambda \mathrm{Tr}(C_\mu^+(\lambda)) + 2i\sqrt{\lambda-\mu-m^2/4}\, \mathrm{Tr}((\tilde{D}_M + i\sqrt{\lambda-\mu-m^2/4}\, I)C_\mu^+(\lambda)) \quad .$$

Assume that $m^2/4+\mu \neq 0$ and let λ tend to $m^2/4+\mu$. Then we get $\mathrm{Tr}(C_\mu^+(m^2/4+\mu)) = \mathrm{Tr}(C_\mu^-(m^2/4+\mu))$. Now consider the case when $m^2/4+\mu = 0$. Since $\Delta_M^+ + m^2/4\, I = \tilde{D}_M^2$, we have $\tilde{D}_M \mid A^+(\mu) = 0$. Thus we obtain

$$\mathrm{Tr}(C_\mu^-(\lambda)) = \mathrm{Tr}(C_\mu^+(\lambda)) - 2\mathrm{Tr}(C_\mu^+(\lambda)) = -\mathrm{Tr}(C_\mu^+(\lambda)) \quad .$$

Q.E.D.

Using (10.18) and Corollary 10.20, we obtain

LEMMA 10.21. The function $\xi(\lambda) = \xi_+^c(\lambda;H,\tilde{H}_o) - \xi_-^c(\lambda;H,\tilde{H}_o)$ is continuous on $\mathbb{R}^+$ and satisfies $\xi(\lambda) = 0$ for $\lambda \in \mathbb{R}^+ - \sigma_{ac}(H)$ and

$$\lim_{\lambda \to 0^+} \xi(\lambda) = -\frac{1}{2}\dim(\ker \tilde{D}_M) - \frac{1}{2}\mathrm{Tr}(C^+_{-m^2/4}(0))$$

if $\sigma_{ac}(H) = \mathbb{R}^+$.

Next we shall employ the fact that H and $\tilde{H}_o$ are supersymmetric Hamiltonians. In fact, the operators Q, Q_o together with the involution τ define a supersymmetric scattering theory in the Hilbert space $\mathcal{H} = L^2(X,E)$ in the sense of [82, Definition 2.5]. For all details of the following discussion we refer to [82]. First we claim that the operator

$$Q\exp(-tH) - Q_o\exp(-t\tilde{H}_o)P_o^{ac} \tag{10.22}$$

is of the trace class. This can be established in the same manner as Theorem 9.1. The trace class property of (10.22) implies that the wave operators $W_\pm(H,\tilde{H}_o)$ intertwine Q and Q_o:

$$QW_\pm(H,\tilde{H}_o) = W_\pm(H,\tilde{H}_o)Q_o \quad \text{on } \mathbb{D}(Q_o) \quad .$$

The proof of this fact is given by Lemma 2.6 in [82] . Now consider the scattering operator

$$S(H,\tilde{H}_o) = W_+(H,\tilde{H}_o)^* W_-(H,\tilde{H}_o) .$$

Then we have

$$Q_o S(H,\tilde{H}_o) = S(H,\tilde{H}_o) Q_o \quad \text{on} \quad \mathbb{D}(Q_o) \quad . \tag{10.23}$$

Moreover, $S(H,\tilde{H}_o)$ commutes with the involution τ . Therefore it decomposes as

$$S(H,\tilde{H}_o) = S_+(H,\tilde{H}_o) \oplus S_-(H,\tilde{H}_o) \quad ,$$

where $S_\pm(H,\tilde{H}_o)$ acts on $H^\pm_{o,ac} = L^2(\mathbb{R}^+ \times \Gamma_M \backslash X_M, E^\pm_M)$. Let $dE(\lambda)$ denote the spectral measure of $\tilde{H}_o$. Since $S(H,\tilde{H}_o)$ and Q_o commute with $\tilde{H}_o$ we get spectral decompositions

$$S(H,\tilde{H}_o) = \int S(\lambda;H,\tilde{H}_o)dE(\lambda) \quad , \quad Q_o = \int Q_o(\lambda)dE(\lambda)$$

and (10.23) implies

$$Q_o(\lambda)S(\lambda;H,\tilde{H}_o) = S(\lambda;H,\tilde{H}_o)Q_o(\lambda) . \tag{10.24}$$

Written in matrix form we have

$$Q_o(\lambda) = \begin{pmatrix} 0 & q_-(\lambda) \\ q_+(\lambda) & 0 \end{pmatrix} \quad , \quad S(\lambda;H,\tilde{H}_o) = \begin{pmatrix} S_+(\lambda;H,\tilde{H}_o) & 0 \\ 0 & S_-(\lambda;H,\tilde{H}_o) \end{pmatrix}$$

and (10.24) gives

$$q_+(\lambda)S_+(\lambda;H,\tilde{H}_o) = S_-(\lambda;H,\tilde{H}_o)q_+(\lambda) \quad .$$

In particular, we may conclude that

$$\det S_+(\lambda;H,\tilde{H}_o) = \det S_-(\lambda;H,\tilde{H}_o) \quad . \tag{10.25}$$

On the other hand, by the Krein-Birman theory of spectral shift functions one has

$$\exp(2i\pi\,\xi^c_\pm(\lambda;H,\tilde{H}_o)) = \det S_\pm(\lambda;H,\tilde{H}_o)$$

for $\lambda\in\sigma_{ac}(\tilde{H}_o)$ [16,V,19.1.5]. Employing (10.25), it follows that

$$\xi_+^c(\lambda;H,\tilde{H}_o) - \xi_-^c(\lambda;H,\tilde{H}_o)\in \mathbb{Z}$$

for each $\lambda\in\sigma_{ac}(H_o)$. Together with Lemma 10.21 this leads to

PROPOSITION 10.26. Let $h = \dim(\ker \tilde{D}_M)$. One has

$$\xi_+^c(\lambda;H,\tilde{H}_o) - \xi_-^c(\lambda;H,\tilde{H}_o) = -\frac{h}{2} - \frac{1}{2}\operatorname{Tr}(C^+_{-m^2/4}(0)) .$$

Thus we have completely determined the integral on the right hand side of the index formula (10.15). Now we turn to the first term on the right hand side of (10.15). Let $e^{\pm}(z,z',t)$ be the kernel of the heat operator $\exp(-tH^{\pm})$ and $\tilde{e}_o^{\pm}(z,z',t)$ the kernel of $\exp(-t\tilde{H}_o^{\pm})$. Then

$$e(z,z',t) = \begin{pmatrix} e^+(z,z',t) & 0 \\ 0 & e^-(z,z',t)\end{pmatrix}$$

is the kernel of $\exp(-tH)$. It follows from Theorem 9.1 and Lemma 10.9 that the functions $z\in X\longmapsto \operatorname{tr} e^{\pm}(z,z,t) - \operatorname{tr}\tilde{e}_o^{\pm}(z,z,t)$ are integrable. Furthermore, the kernels $\tilde{e}_o^{\pm}(z,z',t)$ are described on p.52 in [7] and it is proved that $\operatorname{tr}\tilde{e}_o^+((r,x),(r,x),t) - \operatorname{tr}\tilde{e}_o^-((r,x),(r,x),t)$ is integrable over $\mathbb{R}^+ \times \Gamma_M\backslash X_M$ (c.f.(2.22) in [7]). Thus it follows that the function $z\in X\longmapsto \operatorname{tr} e^+(z,z,t) - \operatorname{tr} e^-(z,z,t)$ is integrable and we obtain

$$\begin{aligned}\operatorname{Tr}(\tau(\exp(-tH) - \exp(-t\tilde{H}_o)P_o^{ac}) &= \int_X (\operatorname{tr} e^+(z,z,t) - \operatorname{tr} e^-(z,z,t))dz \; - \\ &- \int_o^\infty \int_{\Gamma_M\backslash X_M} (\operatorname{tr}\tilde{e}_o^+((r,x),(r,x),t) - \operatorname{tr}\tilde{e}_o^-((r,x),(r,x),t))drdx \quad .\end{aligned} \tag{10.27}$$

According to formula (2.23) in [7] , the second integral in (10.27) is equal to

$$K(t) = -\sum_\omega \frac{\operatorname{sign}\omega}{2}\operatorname{erfc}(|\omega|\sqrt{t}) \quad ,$$

where ω runs over the eigenvalues of $\tilde{D}_M$ and erfc is the complementary error function. $K(t)$ is closely related to the Eta function

$$\eta(s) = \sum_{\omega\neq 0} \frac{\operatorname{sign}\omega}{|\omega|^s} \qquad , \operatorname{Re}(s) > \dim X_M \quad ,$$

of the operator $\tilde{D}_M$. By (2.25) in [7] , one has

$$\int_0^\infty (K(t) + \frac{h}{2})t^{s-1}dt = - \frac{\Gamma(s+\frac{1}{2})}{2s\sqrt{\pi}} \eta(2s) .$$

According to [8,Proposition 2.8] , $\eta(s)$ has an analytic continuation to a meromorphic function in the whole s-plane with at most simple poles. Moreover by [35, Theorem 0.1] , $\eta(s)$ is regular at $s=0$. Using the inverse Mellin transform as in [29,p.50] , it follows that $K(t)$ has an asymptotic expansion

$$K(t) \sim \sum_{j \geq -q} a_j t^{j/2} \qquad \text{as} \quad t \to 0^+ , \tag{10.28}$$

and the constant term a_o in this expansion is given by

$$a_o = - \frac{1}{2}(\eta(0) + h) . \tag{10.29}$$

In view of (10.15) and Proposition 10.26, the first integral on the right hand side of (10.27) has also an asymptotic expansion as $t \to 0^+$. Fix $b > 0$ and construct parametrices $P^{\pm}$ with respect to the decomposition $X = X_b \cup Y_{b+1}$ as in §7. Let $Q^{\pm}$ be the kernels defined by (7.6). Using the properties of $Q^{\pm}$ listed in §7, it is easy to see that $\int_X \operatorname{tr} P^{\pm} * Q^{\pm}(z,z,t)dz$ is exponentially small as $t \to 0^+$. Thus, in order to determine the asymptotic expansion of

$$\int_X (\operatorname{tr} e^+(z,z,t) - \operatorname{tr} e^-(z,z,t))dz$$

as $t \to 0^+$, we can replace $e^{\pm}$ by its parametrix $P^{\pm}$. Note that the asymptotic expansion is independent of the choise of $b > 0$. Given $z \in X$, choose $b >> 0$ such that $z \in X_b$. It follows from the construction of the parametrix that $\operatorname{tr} P^{\pm}(z,z,t)$ and therefore $\operatorname{tr} e^{\pm}(z,z,t)$ has an asymptotic expansion as $t \to 0^+$. Denote by $\alpha^{\pm}(z)$ the constant term of this expansion. Our assumption that D is a generalized Dirac operator implies that $\alpha^+(z) - \alpha^-(z)$ is given on Y_1 by a universal polynomial in the components of the metric tensor and its covariant derivatives at the point z. Since X is locally symmetric near infinity it follows that $\alpha^+(z) - \alpha^-(z)$ is a bounded function on X and therefore integrable.

Let Δ^+ (resp. Δ^-) be the unique selfadjoint extension of $\mathcal{D}^*\mathcal{D}$ (resp. $\mathcal{D}\mathcal{D}^*$) acting on $C_c^\infty(Y,E^{\pm})$ and let $e_1^{\pm}(z,z',t)$ be the kernel of the heat operator $\exp(-t\Delta^{\pm})$. Combining the observations above it follows that

$$\int_{Y_b} (\operatorname{tr} e_1^+(z,z,t) - \operatorname{tr} e_1^-(z,z,t))dz \tag{10.30}$$

has an asymptotic expansion as $t \to 0^+$. Let $\mathcal{U}(b)$ be the constant term of this expansion. Set

$$\mathcal{U} = \lim_{b \to \infty} \mathcal{U}(b) . \tag{10.31}$$

It is easy to see that this limit exists. We shall call $\mathcal{U}$ the unipotent contribution to the index. Together with (10.29) it follows that the constant term of the asymptotic expansion of (10.27) is given by

$$\int_X (\alpha^+(z) - \alpha^-(z))dz + \mathcal{U} + \frac{1}{2}(\eta(0)+h) .$$

Using (10.15) and Proposition 10.26 we may summmarize our results by

<u>THEOREM 10.32</u>. Let X be a Riemannian manifold with a cusp of rank one and let

$$D: C_c^\infty(X,E^+) \longrightarrow C_c^\infty(X,E^-)$$

be a generalized chiral Dirac operator with associated locally invariant operator

$$\mathcal{D}: C_c^\infty(Y,\mathcal{E}^+) \longrightarrow C_c^\infty(Y,\mathcal{E}^-) .$$

Introduce the following objects associated to D:

(i) Let $\alpha^\pm(z)$ be the constant term in the asymptotic expansion of $\operatorname{tr} e^\pm(z,z,t)$ where e^+ and e^- are the heat kernels for the selfadjoint extensions of D^*D and DD^*, respectively.

(ii) Let $\mathcal{U}$ be defined by (10.31).

(iii) Let $D_M: C^\infty(\Gamma_M\backslash X_M, \mathcal{E}_M^+) \longrightarrow C^\infty(\Gamma_M\backslash X_M, \mathcal{E}_M^+)$ be the elliptic selfadjoint operator associated to D by Lemma 10.3 and let $\eta(0)$ be the Eta invariant of $\tilde{D}_M = D_M + m^2/4\, I$.

(iv) Set

$$C^+(\lambda) = \begin{cases} 0 & , \text{ if } \ker \tilde{D}_M = 0 \\ C^+_{-m^2/4}(\lambda) & , \text{ if } \ker \tilde{D}_M \neq 0 \end{cases}$$

Then the L^2-index of D is given by

$$L^2\text{-Ind}(D) = \int_X (\alpha^+(z) - \alpha^-(z))dz + \mathcal{U} + \frac{1}{2}\eta(0) - \frac{1}{2}\operatorname{Tr}(C^+(0)).$$

This is only a preliminary version of the index formula we are aiming for. The main problem that remains is to investigate the unipotent contribution $\mathcal{U}$. This will be done in the next section. Another troublesome term in the index formula is the last one which we shall discuss now. This term arises iff $\ker(\tilde{D}_M) \neq 0$. Now recall that $\tilde{D}_M{}^2 = \Delta_M^+ + m^2/4\, I$. Hence $\ker(\tilde{D}_M) \neq 0$ iff $-m^2/4$ is an eigenvalue of Δ_M^+. In this case $\mu_1 = -m^2/4$ is the lowest eigenvalue of Δ_M^+ and Theorem 4.38 combined with Theorem 6.17 shows that the absolutely continuous spectrum of H^+ extends to 0. But this means that $\bar{D}\colon \mathbb{D}(D) \longrightarrow L^2(X,E^-)$ is not a Fredholm operator. The converse is also true. Thus, the last term in the index formula arises iff $\bar{D}$ is not a Fredholm operator. Now assume that $\ker(\tilde{D}_M) \neq 0$ or, equivalently, that the lowest eigenvalue of Δ_M^+ is $\mu_1 = -m^2/4$. By Lemma 10.3, $-m^2/4$ is also the lowest eigenvalue of Δ_M^-. Since $m^2/4 + \mu_1 = 0$ and μ_1 is the lowest eigenvalue of $\Delta_M^{\pm}$ the automorphism γ_{μ_1} of the spectral surface Σ associated to $\Delta_M^{\pm}$ in §8 is simply given by $\gamma_{\mu_1}(\Lambda) = -\Lambda$. Thus, by Theorem 8.14,(i), $C_{\mu_1}^{\pm}$ satisfies the functional equation

$$C_{\mu_1}^{\pm}(-\Lambda)C_{\mu_1}^{\pm}(\Lambda) = \mathrm{Id}\ .$$

In particular we have $C_{\mu_1}^{\pm}(0)^2 = \mathrm{Id}$ and therefore, each eigenvalue of $C_{\mu_1}^{\pm}(0)$ equals either +1 or -1. Further, recall that $A^+(\mu_1) = \ker(\tilde{D}_M)$. Hence we may regard $C_{\mu_1}^{\pm}(0)$ as acting on $\ker(\tilde{D}_M)$. Suppose that $\Phi \in \ker(\tilde{D}_M)$ is a nonzero eigenvector of $C_{\mu_1}^{\pm}(0)$ with eigenvalue 1 and consider the Eisenstein function $E_{\mu_1}^+(\Phi,\Lambda)$ associated to Φ . Note that the equation $C_{\mu_1}^{\pm}(0)^2 = \mathrm{Id}$ also implies that $C_{\mu_1}^{\pm}(\Lambda)$ is regular at $\Lambda = 0$. Now apply Proposition 9.17 to $E_{\mu_1}^+(\Phi,\tau)$ with $\pi = \mu_1$. Since the left hand side of the equation in Proposition 9.17 is non negative, it follows that each operator $T_{\mu\mu_1}^+(\Lambda)$ is regular at $\Lambda = 0$. Thus the right hand side of the Maaß-Selberg relation for $E_{\mu_1}^+(\Phi,\Lambda)$ is finite at $\Lambda = 0$ and this implies that $E_{\mu_1}^+(\Phi,\Lambda)$ is regular at $\Lambda = 0$. In the same manner one can prove that $E_{\mu_1}^-(\eta\Phi,\Lambda)$ is regular at $\Lambda=0$. Let $\lambda \in T$ Since $\tilde{D}_M\Phi = 0$ it follows from the equation derived in the proof of Lemma 10.19 that

$$DE_{\mu_1}^+(\Phi,\lambda) = i\sqrt{\lambda}E_{\mu_1}^-(\eta\Phi,\lambda)\ .$$

Passing to the limit $\lambda \longrightarrow 0$ we obtain

$$DE_{\mu_1}^+(\Phi,0) = 0\ .$$

Now consider the constant term $(E^+_{\mu_1}(\Phi,0))_o$ of $E^+_{\mu_1}(\Phi,0)$. Employing Lemma 8.10, it follows that

$$(E^+_{\mu_1}(\Phi,0))_o = 2r^{m/2}\Phi + \Psi_o \quad ,$$

where $\Psi_o \in L^2$. On the other hand, if we apply Proposition 9.17 to the truncated Eisenstein function $E^{+,b}_{\mu_1}(\Phi,\tau)$ and pass to the limit $\tau \to 0$ then the discussion above shows that $E^{+,b}_{\mu_1}(\Phi,0)$ is square integrable. Hence, on Y_1 we can write

$$E^+_{\mu_1}(\Phi,0) = 2r^{m/2}\Phi + \Psi$$

with $\Psi \in L^2(Y_1, E^+|Y_1)$. There is a close anology with the extended L^2-sections investigated in [7,p. 58]. For this reason we shall call $E^+_{\mu_1}(\Phi,0)$ an extended L^2-solution of D with limiting value $\Phi \in \ker(\tilde{D}_M)$.

Now assume that $\Phi \in \ker(\tilde{D}_M)$ satisfies $C^+_{\mu_1}(0)\Phi = -\Phi$. Employing Lemma 10.19, it follows that $C^-_{\mu_1}(0)\eta\Phi = \eta\Phi$ and proceeding in the same way as before it follows that $E^-_{\mu_1}(\eta\Phi,\Lambda)$ is regular at $\Lambda=0$ and the section $E^-_{\mu_1}(\eta\Phi,0)$ satisfies

$$D^* E^-_{\mu_1}(\eta\Phi,0) = 0$$

$$E^-_{\mu_1}(\eta\Phi,0) = 2r^{m/2}\eta\Phi + \Psi' \qquad \text{on } Y_1$$

with $\Psi' \in L^2(Y_1, E^-|Y_1)$. We call $E^-_{\mu_1}(\eta\Phi,0)$ an extended L^2-solution of D^* with limiting value $\Phi \in \ker(\tilde{D}_M)$. Let $L^\pm$ denote the ± 1 eigenspaces of $C^+_{\mu_1}(0)$ acting on $\ker(\tilde{D}_M)$. Then we have an orthogonal sum decomposition

$$\ker(\tilde{D}_M) = L^+ \oplus L^-$$

and each element of L^+ (resp. L^-) is the limiting value of an extended L^2-solution of D (resp. D^*). Put

$$h^\pm_\infty = \dim L^\pm \quad .$$

Then the last term in the index formula of Theorem 10.32 can be re-written as

$$\frac{1}{2}(h^+_\infty - h^-_\infty) \quad . \tag{10.33}$$

We may also consider the subspaces $\tilde{L}^+, \tilde{L}^- \subset \ker(\tilde{D}_M)$ consisting of limiting values of all extended L^2-solutions of D and D^*, respectively. More precisely, by an extended L^2-section of E^+ we shall mean a section φ of E^+ which is locally in L^2 and such that, on Y_1, it can be written as $\varphi = r^{m/2}\Phi + \Psi$ where Ψ is in L^2 and Φ $\ker(\tilde{D}_M)$. In anology with [7] we shall call Φ the limiting value of φ. A similar definition holds for extended L^2-sections of E^- (using the isomorphism η). Let $\tilde{L}^+$ and $\tilde{L}^-$ denote the subspaces of $\ker(\tilde{D}_M)$ consisting of the limiting values of all extended L^2-solutions of D and D*, respectively. Then $L^{\pm} \subset \tilde{L}^{\pm}$ and it is very likely - although we shall not prove it here - that $L^{\pm} = \tilde{L}^{\pm}$. But even in this case we do not know how to compute (10.33). In our applications we will have $\ker(\tilde{D}_M) = 0$ so that this term does not occure.

We conclude this chapter by the remark that the index formula can be easily extended to the case of a manifold with several cusps.

CHAPTER XI

THE UNIPOTENT CONTRIBUTION TO THE INDEX

In this chapter we shall investigate the unipotent contribution $\mathcal{U}$ to the L^2-index of D. The main result will an expression for $\mathcal{U}$ in terms of certain unipotent orbital integrals and a noninvariant integral.

Let the notations be the same as in Ch.X. We introduce some additional notation. Let T be a maximal torus of K with Lie algebra $\mathfrak{t}$. Let $\Phi_c = \Phi(\mathfrak{k}_{\mathbb{C}}, \mathfrak{t}_{\mathbb{C}})$ be the root system of $\mathfrak{k}_{\mathbb{C}}$ with respect to $\mathfrak{t}_{\mathbb{C}}$. We fix once and for all a positive system of compact roots $\Psi_c \subset \Phi_c$ and denote

$$\rho_c = \frac{1}{2} \sum_{\alpha \in \Psi_c} \alpha \quad .$$

Under the assumption that $\operatorname{rank} G = \operatorname{rank} K$ we let $\Phi = \Phi(\mathfrak{g}_{\mathbb{C}}, \mathfrak{t}_{\mathbb{C}})$ be the root system of $\mathfrak{g}_{\mathbb{C}}$ with respect to $\mathfrak{t}_{\mathbb{C}}$ and set $\Phi_n = \Phi - \Phi_c$.

Let $\mu \in \mathfrak{t}_{\mathbb{C}}^*$ be the highest weight of an irreducible $\mathfrak{k}_{\mathbb{C}}$-module V_μ. Then we also fix a positive root system $\Psi_\mu \subset \Phi$ such that

$$\mu + \rho_c \quad \text{is} \quad \Psi_\mu\text{-dominant.}$$

Since $\mu+\rho_c$ is Φ_c-regular, it follows that $\Psi_c \subset \Psi_\mu$. Let

$$\rho_\mu = \frac{1}{2} \sum_{\alpha \in \Psi_\mu} \alpha \quad .$$

If no misunderstanding can arise, we shall suppress the subscript μ in the notation.

Let $h_t^{\pm} : G \longrightarrow \operatorname{End}(V^{\pm})$ be the function representing the kernel of $\exp(-t\tilde{\Delta}^{\pm})$. Then

$$e_1^{\pm}(g,g',t) = \sum_{\gamma \in \Gamma} h_t^{\pm}(g^{-1}\gamma g') \tag{11.1}$$

is the kernel of $\exp(-t\Delta^{\pm})$. Put

$$e_o^{\pm}(g,g',t) = \int_{\Gamma \cap U \backslash U} e_1^{\pm}(ug,g',t)du \quad .$$

We may identify $e_o^{\pm}$ with a C^∞ section of $E_M^{\pm} \boxtimes E_M^{\pm}$ considered as a vector bundle over $(\mathbb{R}^+ \times \Gamma_M \backslash X_M) \times (\mathbb{R}^+ \times \Gamma_M \backslash X_M) \times \mathbb{R}^+$ and, similarly to

(9.4), we obtain

$$e_o^{\pm}((r,x),(r,x),t) = \frac{(rr')^{m/2}}{\sqrt{4\pi t}} \exp(-\frac{m^2}{4}t - \frac{\log^2(r/r')}{4t})p^{\pm}(x,x',t)$$

where $p^{\pm}(x,x',t)$ is the heat kernel for $\Delta_M^{\pm}$. In view of Lemma 10.3, one has $\operatorname{tr} p^{+}(x,x,t) = \operatorname{tr} e^{-}(x,x,t)$. Thus we get

$$\operatorname{tr} e_o^{+}((r,x),(r,x),t) = \operatorname{tr} e_o^{-}((r,x),(r,x),t) . \tag{11.2}$$

Set

$$f_t(g) = \operatorname{tr} h_t^{+}(g) - \operatorname{tr} h_t^{-}(g) \quad , \quad g\in G \ , \ t > 0 . \tag{11.3}$$

f_t is K-finite and, according to Proposition 3.16, it belongs to $C^p(G)$ for all $p > 0$. Observe that f_t satisfies

$$f_t(kgk^{-1}) = f_t(g) \quad , \quad k\in K .$$

For $b > 0$, let G_b be defined by (4.10). Employing (10.2), it follows that (10.30) is equal to

$$\int_{\Gamma\backslash G_b} \sum_{\gamma\in\Gamma} (f_t(g^{-1}\gamma g) - \int_{\Gamma\cap U\backslash U} f_t(g^{-1}u\gamma g)du)dg \quad . \tag{11.4}$$

To study the asymptotic behaviour (as $t\to 0$) of (11.4) we decompose Γ. Recall that $\Gamma_M = U\Gamma \cap M$ and $U\cap M = \{1\}$. This implies that

$$\Gamma = \bigsqcup_{\delta\in\Gamma_M} (U\delta \cap \Gamma) \quad . \tag{11.5}$$

Let $\delta\in\Gamma_M$ be fixed and consider the partial sum

$$\sum_{\gamma\in U\delta\cap\Gamma} (f_t(g^{-1}\gamma g) - \int_{\Gamma\cap U\backslash U} f_t(g^{-1}u\gamma g)du) \quad . \tag{11.6}$$

We shall now estimate this sum. To begin with, recall that the Lie algebra $\mathfrak{u}$ of U admits the direct sum decomposition $\mathfrak{u} = \mathfrak{u}_\lambda \oplus \mathfrak{u}_{2\lambda}$, λ being the unique simple root of (P,A). Let

$$U_\lambda = \exp(\mathfrak{u}_\lambda) \quad , \quad U_{2\lambda} = \exp(\mathfrak{u}_{2\lambda}) \quad ,$$

so that $U_\lambda = U_\lambda \cdot U_{2\lambda}$ with $U_\lambda \cap U_{2\lambda} = \{1\}$. If U is non abelian, then U is a two-step nilpotent group with center $U_{2\lambda}$. Put

$$\Gamma_{2\lambda} = \Gamma \cap U_{2\lambda} .$$

Then $\log(\Gamma_{2\lambda})$ is a lattice in $\mathfrak{u}_{2\lambda}$ which we denote by $L_{2\lambda}$. Further, set

$$\Gamma_{\lambda} = \Gamma \cap U/\Gamma_{2\lambda} .$$

Then, with respect to the identification $U_\lambda \simeq U/U_{2\lambda}$, we have $\Gamma_\lambda \subset U_\lambda$, and so $\log(\Gamma_\lambda)$ is a lattice in $\mathfrak{u}_\lambda$ which we call L_λ. For the following computations we shall normalize the invariant measures on U_λ and $U_{2\lambda}$ by demanding that $\mathrm{Vol}(\Gamma_\lambda \backslash U_\lambda) = 1$ and $\mathrm{Vol}(\Gamma_{2\lambda} \backslash U_{2\lambda}) = 1$. $(.,.)_\Theta$ induces a Euclidean structure on $\mathfrak{u}_\lambda$ and $\mathfrak{u}_{2\lambda}$. For $\hat{Z} \in \mathfrak{u}_{2\lambda}$ and $g, g' \in G$ set

$$F_t(g,g',\hat{Z}) = \int_{\mathfrak{u}_{2\lambda}} f_t(g^{-1}\exp(Z)g')e^{2\pi i\langle Z,\hat{Z}\rangle}dZ \quad . \tag{11.7}$$

We observe that $f_t(g^{-1}\exp(Z)g')$, considered as a function of $Z \in \mathfrak{u}_{2\lambda}$ is rapidely decreasing. Let $L^*_{2\lambda} \subset \mathfrak{u}_{2\lambda}$ be the dual lattice of $L_{2\lambda}$. Choose $\gamma_o \in \Gamma \cap (U\delta)$. Then it is clear that

$$U\delta \cap \Gamma = (\Gamma \cap U)\gamma_o \quad .$$

Applying the Poisson summation formula, we obtain

$$\begin{aligned} \sum_{\gamma \in U\delta \cap \Gamma} f_t(g^{-1}\gamma g') &= \sum_{\gamma_1 \in \Gamma_\lambda} \sum_{\gamma_2 \in \Gamma_{2\lambda}} f_t(g^{-1}\gamma_2\gamma_1\gamma_o g') = \\ &= \sum_{\gamma_1 \in \Gamma_\lambda} \sum_{l \in L^*_{2\lambda}} F_t(g,\gamma_1\gamma_o g',l) \quad . \end{aligned} \tag{11.8}$$

Observe that F_t satisfies $F_t(gk,g'k',\hat{Z}) = F_t(g,g',\hat{Z})$ for all $k,k' \in K$, $g,g' \in G$. Therefore, in order to estimate $F_t(g,g',\hat{Z})$, we can assume that $g,g' \in P$. Let $Z_1,\ldots,Z_\nu$ be an orthonormal basis for $\mathfrak{u}_{2\lambda}$ with respect to $(.,.)_\Theta$. Set

$$\Delta_{2\lambda} = \sum_{j=1}^{\nu} L(Z_j)^2 \quad ,$$

where L denotes the left regular representation. Assume that $\hat{Z} \neq 0$ and let $p,p' \in P$. Recall that P normalizes U. If we integrate by parts, then we get

$$F_t(p,p',\hat{Z}) = |\det(\mathrm{Ad}(p)|\mathfrak{u}_{2\lambda})| \int_{\mathfrak{u}_{2\lambda}} f_t(\exp(Z)p^{-1}p')e^{2\pi i\langle Z,\mathrm{Ad}(p)^*Z\rangle}dZ = \tag{11.9}$$

$$= |\det(\mathrm{Ad}(p)|\mathfrak{u}_{2\lambda})|(4\pi^2\|\mathrm{Ad}(p)^*\hat{Z}\|^2)^{-\mu}\int_{\mathfrak{u}_{2\lambda}} ((\Delta_{2\lambda})^\mu f_t)(\exp(Z)p^{-1}p')\cdot$$

$$\cdot e^{2\pi i\langle Z,\mathrm{Ad}(p)^*\hat{Z}\rangle}dZ ,$$

for each $\mu\in\mathbb{N}$.

Let $g\in G$ and assume that $g = pak$ with $p\in UM$, $a\in A$, $k\in K$. Then we set

$$H(g) = \log a .$$

According to the definition of a split component of P (c.f.[66,p.32]) one has $\det(\mathrm{Ad}(m)|\mathfrak{u}_{2\lambda}) = 1$ for all $m\in M$. Thus we get

$$|\det(\mathrm{Ad}(p)|\mathfrak{u}_{2\lambda})| = e^{2m(2\lambda)\lambda(H(p))} , \quad p\in P .$$

Let $\omega \subset UM$ be a compact subset. There exists a constant $C > 0$ such that $\|\mathrm{Ad}(p)^*\hat{Z}\| \geq Ce^{2\lambda(H(p))}\|\hat{Z}\|$, for $\hat{Z}\in\mathfrak{u}_{2\lambda}$ and $p\in\omega A_1$, where A_1 is given by (2.7). To estimate the integral on the right hand side of (11.9) we shall apply formula (3.17). Recall that $h_t^\pm$ can be identified with a section $E^\pm(x,y,t)$ of $E^\pm\boxtimes(E^\pm)^*$ over $G/K\times G/K$. Let $x_o\in G/K$ be the coset of the identity. Then one has $E^\pm(x_o,gx_o,t) = [1\times g,h_t(g)]$. Set

$$\tilde{\Delta}_{2\lambda} = \sum_{j=1}^{\nu} \nabla^2_{Z_j} .$$

In view of the identification just described, it is clear that

$$\|(\Delta_{2\lambda})^\mu h_t^\pm(g)\| = \|(\tilde{\Delta}_{2\lambda})^\mu_{x_o} E^\pm(x_o,gx_o,t)\| , \tag{11.10}$$

$g\in G$, $\mu\in\mathbb{N}$. On the other hand, by (11.3) we have

$$\|(\Delta_{2\lambda})^\mu f_t(g)\| \leq C(\|(\Delta_{2\lambda})\ h_t^+(g)\| + \|(\Delta_{2\lambda})\ h_t^-(g)\|) \tag{11.11}$$

for a certain constant $C > 0$. Using (11.10) and (11.11) combined with (3.17), it follows from (11.9) that there exist constants $C_1,C_2 > 0$ such that

$$|F_t(p,p',\hat{Z})| \le C_1 e^{(2m(2\lambda)-4\mu)\lambda(H(p))} t^{-n/2-2\mu} \|\hat{Z}\|^{-2\mu} \cdot \tag{11.12}$$

$$\cdot \int_{U_{2\lambda}} \exp\left(-C_2 \frac{d^2(pux_o,p'x_o)}{t}\right) du$$

for $p \in \omega A_1$, $p' \in P$, $\mu \in \mathbb{N}$ and $0 < t \leq 1$. $d(x,y)$ denotes the geodesic distance of $x,y \in G/K$. Since $L^*_{2\lambda}$ is a lattice in $\mathfrak{u}_{2\lambda}$ the series

$$\sum_{l \in L^*_{2\lambda} - \{0\}} \|l\|^{-2\mu}$$

is convergent for $\mu > m(2\lambda)/2$. Moreover, observe that the integral on the right hand side of (11.12) equals

$$e^{-2m(2\lambda)\lambda(H(p))} \int_{\Gamma_{2\lambda}\backslash U_{2\lambda}} \sum_{\gamma \in \Gamma_{2\lambda}} \exp\left(-C_2 \frac{d^2(upx_o,\gamma p'x_o)}{t}\right) du \quad .$$

Assume that $\mu > m(2\lambda)/2$. Then , using (11.12), we obtain

$$\sum_{\gamma_1 \in \Gamma_\lambda} \sum_{l \in L^*_{2\lambda} - \{0\}} |F_t(g,\gamma_1\gamma_o g,l)| \le \tag{11.13}$$

$$\le C_3 e^{-4\mu\lambda(H(g))} t^{-n/2-2\mu} \int_{\Gamma_{2\lambda}\backslash U_{2\lambda}} \sum_{\gamma \in U\delta \cap \Gamma} \exp\left(-C_4 \frac{d^2(ugx_o,\gamma gx_o)}{t}\right) du \quad .$$

for $g \in \omega A_1 K$, $0 < t \leq 1$ and certain constants $C_1, C_2 > 0$.

It remains to investigate

$$\sum_{\gamma_1 \in \Gamma_\lambda} F_t(g,\gamma_1\gamma_o g,0) \quad .$$

For $\hat{Z}' \in \mathfrak{u}_\lambda$ let

$$F^*_t(g,g',\hat{Z}') = \int_{\mathfrak{u}_\lambda} e^{2\pi i\langle Z,\hat{Z}'\rangle} \int_{U_{2\lambda}} f_t(g^{-1}u\exp(Z)g')\,du\,dZ \quad .$$

Using the Poisson summation formula, we get

$$\sum_{\gamma_1 \in \Gamma_\lambda} F_t(g,\gamma_1\gamma_o g,0) = \sum_{l \in L^*_\lambda} F^*_t(g,\gamma_o g,l) \quad . \tag{11.14}$$

In order to estimate $F^*_t(g,g',\hat{Z}')$, $\hat{Z}' \neq 0$, we may repeat the arguments employed in the case of $F_t(g,g',\hat{Z})$. Assume that $\mu > m(\lambda)/2$. Then it follows that there exist constants $C_5, C_6 > 0$ such that

$$\sum_{l\in L_\lambda^*-\{0\}} |F_t^*(g,\gamma_o g,l)| \leq \tag{11.15}$$

$$\leq C_5 e^{-2\mu\lambda(H(g))} t^{-n/2-2\mu} \int_{\Gamma\cap U\backslash U} \sum_{\gamma\in U\delta\cap\Gamma} \exp(-C_6 \frac{d^2(ugx_o,\gamma gx_o)}{t})du \quad ,$$

for $g\in\omega A_1 K$ and $0 < t \leq 1$. Finally, observe that

$$F_t^*(g,\gamma_o g,0) = \int_U f_t(g^{-1}u\gamma_o g)du = \sum_{\gamma\in U\delta\cap\Gamma} \int_{\Gamma\cap U\backslash U} f_t(g^{-1}u\gamma g)du \ . \tag{11.16}$$

Let $U_o \subset U$ be a fundamental domain for $\Gamma\cap U$. Then (11.8) combined with (11.13) - (11.16) implies that (11.6) can be estimated by a constant times

$$e^{-2\mu\lambda(H(g))} t^{-n/2-2\mu} \sup_{u\in U_o} \{ \sum_{\gamma\in U\delta\cap\Gamma} \exp(-C \frac{d^2(ugx_o,\gamma gx_o)}{t}) \}$$

for $g\in\omega A_1 K$, $0 < t \leq 1$ and some constant $C > 0$. Here $\mu\in\mathbb{N}$ is chosen such that $\mu > \max(m(\lambda)/2, m(2\lambda)/2)$. The constants may depend on μ.

Now let $x_o'\in X_M$ be the coset of the identity and denote by $d_M(x,y)$ the geodesic distance of $x,y\in X_M$. Let $\delta\in\Gamma_M$ and assume that $\gamma = u_1\delta\in\Gamma$ for $u_1\in U$. Then, referring to (2.14), it follows that

$$d(umax_o,\gamma u'm'a'x_o) \geq d_M(m,\delta m')$$

for all $u,u'\in U$, $m,m'\in M$ and $a,a'\in A$. Thus, using Lemma 3.20, we obtain

LEMMA 11.17. Let $\omega_1 \subset U$ and $\omega_2 \subset M$ be compact subsets. For each $\mu\in\mathbb{N}$, $\mu > \max(m(\lambda)/2, m(2\lambda)/2)$, there exist constants $C_1, C_2 > 0$ such that, for all $\delta\in\Gamma_M$, one has

$$\left| \sum_{\gamma\in U\delta\cap\Gamma} \{f_t(m^{-1}s^{-1}\gamma sm) - \int_{\Gamma\cap U\backslash U} f_t(m^{-1}s^{-1}u\gamma sm)du\} \right| \leq$$

$$\leq C_1 e^{(m(\lambda)+2m(2\lambda)-2\mu)\lambda(H(s))} \exp(-C_2 \frac{d_M^2(mx_o',\delta mx_o')}{t})$$

for $m\in\omega_2$, $s\in\omega_1 A_1$ and $0 < t \leq 1$.

Since Γ_M is a discrete torsion free subgroup of M it is clear that there exists $\varepsilon > 0$ such that $d_M(mx_o',\delta mx_o') > \varepsilon$ for all $m\in\omega_2$

and $\delta \in \Gamma_M - \{1\}$. Moreover, employing Lemma 4.1 and Lemma 4.2 of [26] in the same way as in [26,p.491] , it follows that there exists a constant $C_3 > 0$ such that

$$\sum_{\delta \in \Gamma_M} \exp(-C_2 \frac{d_M^2(mx_o', \delta m x_o')}{t}) \leq C_3$$

for $m \in \omega_2$ and $0 < t \leq 1$. Consider the function on G which is given by

$$\sum_{\gamma \in \Gamma - (U \cap \Gamma)} (f_t(g^{-1}\gamma g) - \int_{\Gamma \cap U \backslash U} f_t(g^{-1} u \gamma g) du) \quad .$$

Since Γ normalizes $\Gamma \cap U$ it is clear that this function is Γ-invariant. So it can be considered as a function on $\Gamma \backslash G$. Using Lemma 11.17 we get

<u>COROLLARY 11.18</u>. There exist constants $C_4, C_5 > 0$ such that, for $0 < t \leq 1$, one has

$$\left| \int_{\Gamma \backslash G_b} \sum_{\gamma \in \Gamma - (\Gamma \cap U)} \{ f_t(g^{-1}\gamma g) - \int_{\Gamma \cap U \backslash U} f_t(g^{-1} u \gamma g) du \} dg \right| \leq C_4 e^{-C_5/t} \quad ,$$

uniformly for $b \geq 1$.

Now recall that our purpose is to determine the constant term of the asymptotic expansion as $t \to 0$ of (11.4). Thus, in view of Corollary 11.18, we can replace in (11.4) the sum over Γ by the sum over $\Gamma \cap U$. We shall first investigate

$$\sum_{\gamma \in \Gamma \cap U} \int_{\Gamma \cap U \backslash U} f_t(g^{-1} u \gamma g) du = \int_U f_t(g^{-1} u g) du \quad . \tag{11.19}$$

Since U is a connected nilpotent Lie group it is unimodular [47,X, Proposition 1.4] . Therefore, we get

$$\int_U f_t(g^{-1} u g) du = e^{\rho(H(g))} \int_U f_t(u) du \quad .$$

We claim that this term is zero. To prove this we need some preparation.

Let π be a unitary representation of G on a Hilbert space $H(\pi)$. By $H_\infty(\pi)$ we shall denote the subspace of all C^∞-vectors for H . Let

$$Z = \sum X_i \otimes C_i \in (\mathcal{U}(g_{\mathbb{C}}) \otimes \mathrm{Hom}(V^+,V^-))^K$$

be the element which defines the invariant operator

$$\tilde{\mathcal{D}}: C^\infty(G/K,\tilde{E}^+) \longrightarrow C^\infty(G/K,\tilde{E}^-) \quad .$$

Then Z gives rise to an operator $\pi(Z)$ from $\mathcal{H}_\infty(\pi) \otimes V^+$ to $\mathcal{H}_\infty(\pi) \otimes V^-$ by

$$\pi(Z) = \sum \pi(X_i) \otimes C_i \quad .$$

$\pi(Z)$ maps $(\mathcal{H}_\infty(\pi) \otimes V^+)^K$ into $(\mathcal{H}_\infty(\pi) \otimes V^-)^K$ and so it defines an operator from $(\mathcal{H}(\pi) \otimes V^+)^K$ to $(\mathcal{H}(\pi) \otimes V^-)^K$ which we shall denote by $\mathcal{D}_\pi$. Let $Z^* = \sum X_i^* \otimes C_i^*$, where C_i^* is the adjoint of C_i and X_i^* is the image of X_i under the canonical anti-involution of $\mathcal{U}(g_{\mathbb{C}})$. Then $\pi(Z^*)$ is the formal adjoint operator of $\pi(Z)$. Let $\mathcal{D}_\pi^*$ be the restriction of $\pi(Z^*)$ to $(\mathcal{H}_\infty(\pi) \otimes V^-)^K$. Since Z is elliptic it follows that the closure of $\mathcal{D}_\pi^*$ coincides with the Hilbert space adjoint of $\mathcal{D}_\pi$ [61,Corollary 1.2]. Let

$$Q_\pi^\pm = \int_K \pi(k) \otimes \sigma^\pm(k)dk \quad .$$

<u>LEMMA 11.20</u>. Let $L^\pm \in \mathrm{End}(V^\pm)$ be associated to $\mathcal{D}$ via (5.4). Then we have

$$\mathcal{D}_\pi^*\mathcal{D}_\pi = Q_\pi^+(-\pi(\Omega) \otimes \mathrm{Id} + \mathrm{Id} \otimes L^+)Q_\pi^+$$

$$\mathcal{D}_\pi\mathcal{D}_\pi^* = Q_\pi^-(-\pi(\Omega) \otimes \mathrm{Id} + \mathrm{Id} \otimes L^-)Q_\pi^- \quad .$$

<u>PROOF</u>. Let L (resp. R) denote the left (resp. right) regular representation of G on $C^\infty(G)$. We identify $\mathcal{H}_\infty(\pi)$ with the subspace ${}^G(C^\infty(G) \hat{\otimes} \mathcal{H}(\pi))$ of $C^\infty(G) \hat{\otimes} \mathcal{H}(\pi)$ consisting of all elements which are invariant under the representation $g \longmapsto L(g) \otimes \pi(g)$ of G. The representation π coincides with the restriction of $R \otimes \mathrm{Id}$ to the space ${}^G(C^\infty(G) \hat{\otimes} \mathcal{H}(\pi))$. Therefore, we get

$$(\mathcal{H}_\infty(\pi) \hat{\otimes} V^\pm)^K \simeq ({}^G(C^\infty(G) \hat{\otimes} \mathcal{H}(\pi)) \otimes V^\pm)^K \simeq {}^G((C^\infty(G) \otimes V^\pm)^K \hat{\otimes} \mathcal{H}(\pi))$$

and it follows from the definition of $\tilde{\mathcal{D}}$ that, with respect to this identification, $\mathcal{D}_\pi$ (resp. $\mathcal{D}_\pi^*$) coincides with the restriction of $\tilde{\mathcal{D}} \otimes \mathrm{Id}$ (resp. $\tilde{\mathcal{D}}^* \otimes \mathrm{Id}$) to the corresponding subspace. Furthermore, observe that $\pi(\Omega)$ is the restriction of $R(\Omega) \otimes \mathrm{Id}$ to $\mathcal{H}_\infty(\pi) = {}^G(C^\infty(G) \hat{\otimes} \mathcal{H}(\pi))$.

Using these observations, the lemma follows from (5.4). Q.E.D.

We can now proceed as in [15,p.161]. Let p_t be the kernel determined by (3.13). According to [70,Theorem 1.4], one has

$$\pi(p_t) = e^{-t\pi(\Delta)} \quad ,$$

where $\Delta = -\Omega + 2\Omega_K$. Therefore, just as in [15,p.160], we obtain

$$\exp(-t\mathcal{D}_\pi^*\mathcal{D}_\pi) = Q_\pi^+(e^{-t\pi(\Delta)} \otimes e^{t(2\sigma^+(\Omega_K) - L^+)})Q_\pi^+$$

$$\exp(-t\mathcal{D}_\pi\mathcal{D}_\pi^*) = Q_\pi^-(e^{-t\pi(\Delta)} \otimes e^{t(2\sigma^-(\Omega_K) - L^-)})Q_\pi^- \quad .$$

Using (3.15) it follows that

$$\pi(h_t^+) = \exp(-t\mathcal{D}_\pi^*\mathcal{D}_\pi) \quad , \quad \pi(h_t^-) = \exp(-t\mathcal{D}_\pi\mathcal{D}_\pi^*) \quad .$$

Now assume that π is irreducible. Then the spaces $(\mathcal{H}(\pi) \otimes V^\pm)^K$ are finite-dimensional. Therefore, $\mathcal{D}_\pi$ has a finite index, $\pi(h_t^\pm)$ are of the trace class and, by abstract operator-theoretic reasons, we get

$$\operatorname{Ind} \mathcal{D}_\pi = \operatorname{Tr} \pi(h_t^+) - \operatorname{Tr} \pi(h_t^-) = \operatorname{Tr} \pi(f_t) \quad . \tag{11.21}$$

On the other hand, since $\dim(\mathcal{H}(\pi) \otimes V^\pm)^K < \infty$, we have

$$\operatorname{Ind} \mathcal{D}_\pi = \dim(\mathcal{H}(\pi) \otimes V^+)^K - \dim(\mathcal{H}(\pi) \otimes V^-)^K \quad . \tag{11.22}$$

Suppose that $\operatorname{rank} G > \operatorname{rank} K$. Then it turns out that V^+ and V^- are isomorphic as K-modules. This fact was established in the proof of (1.2.5.) in [15]. Thus, using (11.21) and (11.22), we obtain

$$\operatorname{Tr} \pi(f_t) = 0 \quad , \quad \text{if} \quad \operatorname{rank} G > \operatorname{rank} K \quad . \tag{11.23}$$

Now assume that $\operatorname{rank} G = \operatorname{rank} K$. Let $S^\pm$ denote the half-spin $k_{\mathbb{C}}$-modules (c.f. [15,p.157]). Since $\widetilde{\mathcal{D}}$ is elliptic it follows from the results of Miatello [59] that there exists a unique virtual $k_{\mathbb{C}}$-module V such that, in the representation ring $R(k_{\mathbb{C}})$ of $k_{\mathbb{C}}$, one has

$$[V^+] - [V^-] = V \otimes ([S^+] - [S^-]) \quad . \tag{11.24}$$

Moreover, $V \otimes S^\pm$ R(K), the representation ring of K. Assume that

$$V = \sum_{\mu} n_{\mu} E_{\mu} \quad ,$$

where μ ranges over a finite subset of $t_{\mathbb{C}}^{*}$, E_{μ} is the corresponding irreducible $k_{\mathbb{C}}$-module with highest weight μ and $n_{\mu} \in \mathbb{Z}$. Then, using (11.21) and (11.22), we obtain

$$\operatorname{Tr} \pi(f_t) = \sum_{\mu} n_{\mu} (\dim (H(\pi) \otimes E_{\mu} \otimes S^{+})^{K} - \dim (H(\pi) \otimes E_{\mu} \otimes S^{-})^{K} . \qquad (11.25)$$

This gives

<u>PROPOSITION 11.26</u>. There are but finitely many $\pi \in \hat{G}$ such that

$$\operatorname{Tr} \pi(f_t) \neq 0 .$$

<u>PROOF</u>. In view of (11.23), we can assume that $\operatorname{rank} G = \operatorname{rank} K$. Let $\pi \in \hat{G}$ and denote by $\chi_{\pi}: \mathfrak{Z}(g_{\mathbb{C}}) \longrightarrow \mathbb{C}$ the infinitesimal character of π. If $\tau \in t_{\mathbb{C}}^{*}$, we denote by χ_{τ} the character of $\mathfrak{Z}(g_{\mathbb{C}})$ associated to τ by the standard procedure. It is proved in [9,p.18] that

$$\dim (H(\pi) \otimes E_{\mu} \otimes S^{+})^{K} - \dim (H(\pi) \otimes E_{\pi} \otimes S^{-})^{K} = 0 , \qquad (11.27)$$

$$\text{unless} \quad \chi_{\bar{\pi}} = \chi_{\mu+\rho_c}$$

where $\bar{\pi}$ denotes the contragredient of π. Moreover, by a result of Harish-Chandra [40], only finitely many classes $\pi \in \hat{G}$ can have a given infinitesimal character (c.f. also [9,p.19]). These observations combined with (11.25) imply that $\operatorname{Tr} \pi(f_t) = 0$ for almost all $\pi \in \hat{G}$. Q.E.D.

Using this proposition, we can now prove

<u>LEMMA 11.28</u>. One has

$$\int_U f_t(u) du = 0 \ .$$

<u>PROOF</u>. Let $P_1 = U_1 A_1 M_1$ be any parabolic subgroup of M with split component A_1. For $\omega \in (\widehat{M_1})_d$ (the discrete series of M_1) and $\nu \in (a_1)^*$ we shall denote by $\Theta_{P_1,\omega,\nu}$ the character of the unitarily induced representation of M associated with (P_1,ω,ν) (c.f. [78,I,§5]). Set

$$\varphi_t(m) = \int_U f_t(um) du \quad , \quad m \in M \ .$$

Then φ_t is a K-finite function in $\mathcal{C}(M)$ and we have to show that $\varphi_t(1) = 0$. In view of Harish-Chandra's Plancherel formula [46,p.175], it is sufficient to prove that $\Theta_{P_1,\omega,\nu}(\varphi_t) = 0$ for all triples (P_1,ω,ν) as above. Let

$$(f_t)_\nu^{(P)}(m) = \int_A \int_U f_t(uam)\exp\{-(i\nu+\rho)(\log a)\}duda \quad , \quad m\in M .$$

Then it is clear that

$$\varphi_t(m) = \int_{\mathfrak{a}^*} (f_t)_\nu^{(P)}(m)d\nu \quad , \quad m\in M. \tag{11.29}$$

Let (P_1,ω,ν_1) be as above. The character $\Theta_{P_1,\omega,\nu_1}$ is a locally integrable function on M which is analytic on M' (= regular elements of M) (c.f. [45,Theorem 11.1]). Then, using Corollary 13.2 of [45] and (11.29), we obtain

$$\begin{aligned}\Theta_{P_1,\omega,\nu_1}(\varphi_t) &= \int_M \Theta_{P_1,\omega,\nu_1}(m)\varphi_t(m)dm = \\ &= \int_{\mathfrak{a}^*} \Theta_{P_1,\omega,\nu_1}((f_t)_\nu^{(P)})\, d\nu \quad .\end{aligned} \tag{11.30}$$

Let $\tilde{\Theta}_\omega$ be the character of $\omega\in(\widehat{M}_1)_d$. Then, by [46,Lemma 21.2] , one has

$$\Theta_{P_1,\omega,\nu_1}((f_t)_\nu^{(P)}) = \tilde{\Theta}_\omega(\{(f_t)_\nu^{(P)}\}_{-\nu_1}^{(P_1)}) \quad , \tag{11.31}$$

where

$$\alpha_{\nu_1}^{(P_1)}(m_1) = \int_{A_1} \int_{U_1} \alpha(u_1a_1m_1)\exp\{-(i\nu_1+\rho)(\log a)\}du_1da_1 \quad ,$$

$m_1\in M_1$, for $\alpha\in \mathcal{C}(M)$ and $\nu_1\in(\mathfrak{a}_1)^*$.

Now recall that there is a one-one correspondence between parabolic subgroups P' of G which are contained in P and parabolic subgroups P_1 of M. If P' corresponds to P_1, then its Langlands decomposition $P' = U'A'M'$ is given by

$$M' = M_1 \quad , \quad A' = A_1A \quad \text{and} \quad U' = U_1U \quad ,$$

(c.f. [45,Lemma 6.1]). For $\nu'\in(\mathfrak{a}')^*$ set

$$(f_t)^{(P')}_{\nu'}(m') = \int_{A'}\int_{U'} f_t(u'a'm')\exp\{-(i\nu'+\rho)(\log a')\}\,du'da' \quad ,$$

$m'\in M'$. Then it is easy to see that

$$(f_t)^{(P')}_{(\nu_1,\nu)} = \{(f_t)^{(P)}_{\nu}\}^{(P_1)}_{\nu_1} \quad .$$

Referring again to [46,Lemma 21.2], it follows from (11.31) that

$$\Theta_{P_1,\omega,\nu_1}((f_t)^{(P)}_{\nu}) = \Theta_{P',\omega,(-\nu_1,\nu)}(f_t) \quad ,$$

where $\Theta_{P',\omega,(-\nu_1,\nu)}$ is the character of the unitarily induced representation of G associated to $(P',\omega,(-\nu_1,\nu))$. Insert this in (11.30) and then apply Proposition 11.26. Since $\dim \mathfrak{a}^* = 1$ it follows that $\Theta_{P_1,\omega,\nu_1}(\varphi_t) = 0$. Q.E.D.

Now consider the function

$$\sum_{\gamma\in\Gamma\cap U} f_t(g^{-1}\gamma g)$$

of $g\in G$. Since Γ normalizes $\Gamma\cap U$ this function is Γ-invariant and it follows from Lemma 11.17, (11.19) and Lemma 11.28 that it is integrable over $\Gamma\backslash G_b$. Befor investigating this integral we shall separate the contribution given by $1\in\Gamma\cap U$. This contribution is obviously equal to

$$\mathrm{Vol}(\Gamma\backslash G_b)f_t(1) = \mathrm{Vol}(Y_b)f_t(1) \quad . \tag{11.32}$$

One can use Harish-Chandra's Plancherel formula [46,Theorem 27.3] to determine $f_t(1)$. The Plancherel formula together with Proposition 11.26 gives

$$f_t(1) = \sum_{\omega\in\hat{G}_d} d(\omega)\,\Theta_\omega(f_t) \quad , \tag{11.33}$$

where $d(\omega)$ is the formal degree of $\omega\in\hat{G}_d$ and the sum is finite. $\Theta_\omega(f_t)$ can be computed using (11.25). Let $V = \sum n_\mu E_\mu$ be the unique virtual $\mathfrak{k}_{\mathbb{C}}$-module determined by (11.24). For $\mu\in\mathfrak{t}^*_{\mathbb{C}}$, let $\hat{G}_\mu \subset \hat{G}$ be the subspace of all $\pi\in\hat{G}$ whose infinitesimal character is $\chi_{\mu+\rho_c}$. Then, in view of (11.25), (11.27) and (11.32), we obtain

$$f_t(1) = \sum_{\mu} n_{\mu} \sum_{\omega\in\hat{G}_{\mu}\cap\hat{G}_d} d(\omega)\{\dim(H(\omega)\otimes E_{\mu}\otimes S^+)^K - \\ - \dim(H(\omega)\otimes E_{\mu}\otimes S^-)^K\}. \tag{11.34}$$

Thus, $f_t(1)$ is independent of t. Since $\mathrm{Vol}(\Gamma\backslash G_b) \to 0$ as $b\to\infty$, it follows from (10.31) that the term (11.32) gives no contribution to U and $U = \lim_{b\to\infty} U(b)$, where $U(b)$ is the constant term in the asymptotic expansion (as $t \to 0$) of

$$\int_{\Gamma\backslash G_b} \sum_{\gamma\in(\Gamma\cap U)-\{1\}} f_t(g^{-1}\gamma g)\,dg \quad . \tag{11.35}$$

As we shall see now, one can replace $\Gamma\backslash G_b$ by $\Gamma\backslash G$ and, up to a term which is exponentially small as $t \to 0$, the resulting integral coincides with the given one. For this purpose consider the function

$$\Phi_t(a) = \int_{\Gamma\backslash S} \sum_{\gamma\in(\Gamma\cap U)-\{1\}} |f_t(a^{-1}s^{-1}\gamma sa)|\,ds \quad , \quad a\in A .$$

Recall that $S = UM$. Applying (11.10) and (11.11) with $\mu = 0$ together with (3.17), we obtain

$$\Phi_t(a) \le Ct^{-n/2} \int_{\Gamma\backslash S} \sum_{\gamma\in(\Gamma\cap U)-\{1\}} \exp\left(-\frac{d^2(sax_o,\gamma sax_o)}{4t}\right) ds \tag{11.36}$$

for some constant $C > 0$.

LEMMA 11.37. Let $S_o \subset S$ be compact. There exists a constant $C > 0$ such that

$$\min_{\gamma\in\Gamma\cap U-\{1\}} \{d(sax_o,\gamma sax_o)\} \ge C(|\lambda(\log a)| + 1) \quad ,$$

uniformly for $s\in S_o$ and $\lambda(\log a) < 0$.

PROOF. The manifold $\tilde{Y} = G/K$ is simply connected and has non-positive curvature. Such a manifold admits a natural compactification $cp(\tilde{Y}) = \tilde{Y} \cup \tilde{Y}(\infty)$ where the points at infinity are classes of positively asymptotic geodesics (c.f. [53,§3.8]). Two geodesics c, c' on $\tilde{Y}$ are called positively asymptotic if there exists a constant $\eta = \eta(c,c') \ge 0$ such that $d(c(t),c'(t)) \le \eta$ for all $t \ge 0$. For a given geodesic c we shall denote by $c(\infty)$ (resp. $c(-\infty)$) the class of positively asymptotic geodesics determined by $c(t)$ (resp. $c(-t)$), $t \ge 0$. Now recall that A can be identified with $\mathbb{R}$ by sending $a \longrightarrow \lambda(\log a)$. We shall write a_t for the element of A which corresponds to $t \in \mathbb{R}$. Let $s \in S$ and consider the curve $c_s: \mathbb{R} \longrightarrow \tilde{Y}$ given by $c_s(t) = sa_t x_o$. c_s is a geodesic in $\tilde{Y}$ [54,Vol.II,Theorem 3.2 and Theorem 3.3] . Let $s=um$, $s'=u'm'$, where $u,u' \in U$, $m \in M$, and assume that $c_s(-\infty)=c_{s'}(-\infty)$. Using (2.14) it follows that $d(c_s(t),c_{s'}(t)) \longrightarrow 0$ as $t \to \infty$, i.e. $c_s(\infty) = c_{s'}(\infty)$. Then, by Lemma 3.8.6 in [53,p.353] , there exists an isometric totally geodesic immersion

$$F: [0,r] \times \mathbb{R} \longrightarrow \tilde{Y}$$

of the flat strip $[0,r] \times \mathbb{R} \subset \mathbb{R} \times \mathbb{R}$, $r \geq 0$, into $\tilde{Y}$ such that $F|\{0\} = c_s$ and $F|\{r\} = c_{s'}$. Since $d(c_s(t),c_{s'}(t)) \longrightarrow 0$ as $t \to \infty$, it follows that $c_s = c_{s'}$. Therefore $u=u'$. Now let $S_o \subset S$ be compact and $\gamma \in \Gamma \cap U - \{1\}$. Using Lemma 3.8.5 and Proposition 3.8.10 of [53] it follows that there exist $\varepsilon > 0$ and $t_o \geq 0$ such that, for each $s \in S_o$, the angle between the geodesic c_s and the geodesic from sx_o to $c_{\gamma s}(-t)$ is bounded away from zero by ε for all $t \geq t_o$. Using the cosine inequality [53,Lemma 3.8.2] it follows that there exists a constant $C_1 > 0$ such that

$$d(c_s(0),c_s(-t))\, d(c_s(0),c_{\gamma s}(-t)) \leq C_1 d^2(c_s(-t),c_{\gamma s}(-t))$$

for all $s \in S_o$ and $t \geq t_o$. By (2.14), we have $d(c_s(0),c_s(-t)) = t$. Moreover, using again the cosine inequality for the triangle determined by $c_s(0)$, $c_{\gamma s}(0)$ and $c_{\gamma s}(-t)$ we obtain the existence of another constant $C_2 > 0$ such that $d(c_s(0),c_{\gamma s}(-t)) \geq C_2 t$ for $s \in S_o$ and $t \geq t_o$. This implies that $d(c_s(-t),c_{\gamma s}(-t)) \geq C_3 t$, for $s \in S_o$ and $t \geq t_o$. Now let $U_o \subset U$, $M_o \subset M$ be compact sets such that $S_o \subset U_o M_o$ and U_o contains a fundamental domain of $\Gamma \cap U$. Then the minimum of $\{d(sa_t x_o, \gamma sa_t x_o) \mid s \in S_o,\ t \geq t_o,\ \gamma \in \Gamma \cap U - \{1\}\}$ is attained at those $\gamma \in \Gamma \cap U - \{1\}$ for which $\gamma U_o \cap U_o \neq \emptyset$. This set is finite. Therefore, the lemma follows from the estimations above. Q.E.D.

Using Lemma 11.37, (3.21) and the estimation of $\operatorname{diam}(B(z))$ given at the end of Ch.3 it follows that there exist constants $C_1, C_2, C_3 > 0$ and $p \in \mathbb{N}$ such that

$$\sum_{\gamma \in (\Gamma \cap U) - \{1\}} \exp\left(-\frac{d^2(sax_o, \gamma sax_o)}{4t}\right) \leq C_1 a^{p\lambda} e^{-C_2/t} \cdot \exp(-C_3(\lambda(\log a))^2) \tag{11.38}$$

uniformly for $s \in S_o$, $\lambda(\log a) < 0$ and $0 < t \leq 1$.

By (11.36) and (11.38), we obtain

$$\int_{A - A_b} \Phi_t(a)a^{-2\rho}da \leq C_4 e^{-C_2/t}$$

for $0 < t \leq 1$ and some constant C_4. Thus it is justified to replace (11.35) by

$$\int_{\Gamma\backslash G} \sum_{\gamma\in(\Gamma\cap U)-\{1\}} f_t(g^{-1}\gamma g)\, dg \quad , \tag{11.39}$$

or, equivalently, by

$$\int_A \int_{\Gamma_M\backslash M} \int_{\Gamma\cap U\backslash U} \left(\sum_{\gamma\in(\Gamma\cap U)-\{1\}} f_t(a^{-1}m^{-1}u^{-1}\gamma uma) \right) du dm\, a^{-2\rho}da \quad ,$$

and $\mathcal{U}$ is the constant term in the asymptotic expansion (as $t \to 0$) of this integral. For the study of this integral it is convenient to introduce a complex parameter s at the A-level. Set

$$\mathcal{U}(f_t,s) =$$

$$= \int_A \int_{\Gamma_M\backslash M} \int_{\Gamma\cap U\backslash U} \left(\sum_{\gamma\in(\Gamma\cap U)-\{1\}} f_t(a^{-1}m^{-1}u^{-1}\gamma uma) \right) du dm\, a^{-(s\lambda+2\rho)}da \quad ,$$

$s\in\mathbb{C}$.

If we recall the estimations which we used above, then it is clear that $\mathcal{U}(f_t,s)$ is an entire function of s and $\mathcal{U}(f_t,0)$ equals (11.39). Moreover one has

<u>LEMMA 11.40</u>. The integral-series defining $\mathcal{U}(f_t,s)$ is absolutely convergent for $\mathrm{Re}(s) > 0$.

<u>PROOF</u> In view of (11.38), it is clear that

$$\int_{A - A_b} \Phi_t(a)a^{-(\sigma\lambda+2\rho)}da < \infty \quad ,$$

for each $\sigma\in\mathbb{R}$. On the other hand, by (11.36) and Lemma 3.20, we have

$$\Phi_t(a) \leq Ct^{-n/2}a^{2\rho}$$

for some constant $C > 0$ and $0 < t \leq 1$. This shows that

$$\int_{A_b} \Phi_t(a)a^{-(\sigma\lambda+2\rho)}da < \infty$$

for $\sigma > 0$. Q.E.D.

We remark that terms similar to $\mathcal{U}(f_t,s)$ occure in the trace formula for rank one lattices (c.f. [67]). To make this point more explicit consider a rank one lattice Γ in G and assume that $\Gamma\backslash G/K$ has a single cusp. Let R_Γ be the right regular representation of G on $L^2(\Gamma\backslash G)$. There is a direct sum decomposition

$$L^2(\Gamma\backslash G) = L^2_d(\Gamma\backslash G) \oplus L^2_c(\Gamma\backslash G)$$

into invariant subspaces where $L^2_d(\Gamma\backslash G)$ is the direct sum of all irreducible unitary subrepresentations of R_Γ. Let R^d_Γ be the restriction of R_Γ to $L^2_d(\Gamma\backslash G)$. Assume that $\operatorname{rank} G = \operatorname{rank} K$. Let $\mu \in \mathfrak{t}^*_{\mathbb{C}}$ be such that $\mu+\rho_n$ exponentiates to a character of T. Then we can form the twisted Dirac operator $D^\pm_\mu : C^\infty(\Gamma\backslash G/K, \mathcal{E}^\pm_\mu) \longrightarrow C^\infty(\Gamma\backslash G/K, \mathcal{E}^\mp_\mu)$ [15,p.157] and it follows as in [15] that

$$L^2\text{-}\operatorname{Ind} D^\pm_\mu = \operatorname{Tr} R^d_\Gamma(f_{\mu,t}) \quad ,$$

where $f_{\mu,t}$ has the same meaning as above. Note that this trace exists by the results of Donnelly [28] . Now assume that the trace formula, established in [67] under an additional assumption, holds in our situation. Employing this trace formula one may proceed along lines similar to [15] and compute the L^2-index of $D^\pm_\mu$ in this way. Among other terms there appears the parabolic term which, in the terminology of Osborne and Warner, is given by

$$(s) \sum_{\{\delta\}_{\Gamma_M}} \lim_{s\to 0} \frac{d}{ds}(s\, \Psi_{f_{\mu,t}}(\delta:s)) \quad .$$

Here $P = MAU$ is a fixed Γ-cuspidal parabolic subgroup of G, $\Gamma_M = \Gamma U \cap M$ and $\{\delta\}_{\Gamma_M}$ runs over the Γ_M-conjugacy classes of elements of $\Gamma_M(s)$ (= set of all elements of Γ_M which have a non-trivial centralizer in U). For the definition of $\Psi_\alpha(\delta:s)$, $\alpha \in C^p(G)$, see [67, p.79 and p.112]. It turns out that in this case $\Psi_{f_{\mu,t}}(1:s)$ is closely related to $\mathcal{U}(f_{\mu,t},s)$ and the other terms $\Psi_{f_{\mu,t}}(\delta:s)$, $\delta \neq 1$, will make no contribution to the index.

Our problem is now to compute $\mathcal{U}(f_t,s)$. If G is a group of real rank one, then this term can be analysed using the same computations as in [77], [65], [23] and [15]. The point is that, according to a principle of Kostant, M operates transitively on the unit sphere in $\mathfrak{u}_\lambda$ provided that $\dim G > 3$. Moreover, by the same principle, $M \times \mathbb{Z}_2$ operates transitively on the unit sphere in $\mathfrak{u}_{2\lambda}$. This is not true if the real rank of G is greater than one. See [67,§9] for the discussion of a special case. It turns out that the case of Hilbert modular groups, discussed in [63], reflects the general picture. Let us recall some facts of this paper.

Hilbert modular groups are related to $G = (SL(2,\mathbb{R}))^n$, $n>1$. Let Γ be a rank one lattice in G with a single parabolic orbit. Let $P = UAM$ be a Γ-cuspidal parabolic subgroup of G. Then M can be identified with $\Lambda = \{\lambda \in \mathbb{R}^n \mid \lambda_1 \cdots \lambda_n = \pm 1\}$ and $U \simeq \mathbb{R}^n$. The action of M on U is given by $\lambda \cdot x = (\lambda_1^2 x_1, \ldots, \lambda_n^2 x_n)$. Let $\mathring{U} \subset U$ be the subset given by $x_1 \cdots x_n \neq 0$. Then $\mathring{U}$ is the disjoint union of a finite number of AM-orbits. The orbits are parametrized by $\varepsilon \in \{\pm 1\}^n$, where the orbit $\mathring{U}_\varepsilon$ which corresponds to $\varepsilon \in \{\pm 1\}^n$ is given by $\mathring{U}_\varepsilon = \{x \in U \mid \varepsilon_i x_i > 0\}$ Moreover, one has $\Gamma \cap U \subset \mathring{U}$. For $x \in \mathbb{R}^n$, put $N(x) = x_1 \cdots x_n$. To each orbit $\mathring{U}_\varepsilon$ there is associated a zeta function

$$\zeta_\varepsilon(s) = \sum_{\mu \in (\Gamma \cap \mathring{U}_\varepsilon)/\Gamma_M} |N(\mu)|^{-s} , \quad \mathrm{Re}(s) > 1 .$$

Here $\Gamma_M = \Gamma U \cap M$ acts via inner automorphisms on $\Gamma \cap \mathring{U}_\varepsilon$. Each zeta function $\zeta_\varepsilon(s)$ has a meromorphic continuation to the whole complex plane. Furthermore, for each $\varepsilon \in \{\pm 1\}^n$, let $u(\varepsilon) \in U$ be the element corresponding to ε under the identification $U \simeq \mathbb{R}^n$. Then, according to [63,p.84], one has

$$\mathcal{U}(f_t,s) =$$

$$= \mathrm{Vol}(\Gamma \cap U \backslash U) \sum_{\varepsilon \in \{\pm 1\}^n} \zeta_\varepsilon(s) \int_A \int_M f_t(a^{-1}m^{-1}u(\varepsilon)ma)\,dm\,a^{-(s\lambda+2\rho)}\,da .$$

In a recent paper W.Hoffmann [51] has shown that, for an arbitrary rank one lattice in a semi-simple Lie group, one has in principle the same picture. The final answer depends on the orbit structure of the action of $L = AM$ on U via conjugation. His method can be easily extended to our case. In passing we briefly recall some of his results.

We introduce the function

$$I(f_t,g) = \int_{\Gamma\cap U\backslash U} \sum_{\gamma\in(\Gamma\cap U)-\{1\}} f_t(g^{-1}u^{-1}\gamma ug)\,du \quad , \quad g\in G \ .$$

Then we have

$$\mathcal{U}(f_t,s) = \int_A \int_{\Gamma_M\backslash M} I(f_t,ma)\,dm\, a^{-(s\lambda+2\rho)}\,da \quad . \tag{11.41}$$

If we proceed as in [23, §1], then we obtain

$$I(f_t,g) = \mathrm{Vol}(\Gamma\cap U\backslash U)\Big\{ \sum_{\gamma\in\Gamma_{2\lambda}-\{1\}} f_t(g^{-1}\gamma g) \ +$$

$$+ \ \mathrm{Vol}(\Gamma_{2\lambda}\backslash U_{2\lambda})^{-1} \sum_{\gamma\in\Gamma_\lambda-\{1\}} \int_{U_{2\lambda}} f_t(g^{-1}u\gamma g)\,du \Big\} \quad .$$

Now observe that the decomposition $u = u_\lambda \oplus u_{2\lambda}$ is invariant under the adjoint action of M on u. Therefore, Γ_λ and $\Gamma_{2\lambda}$ are invariant under the action of Γ_M on U_λ and $U_{2\lambda}$, respectively, by inner automorphisms. For each $\gamma\in\Gamma_\lambda$ or $\gamma\in\Gamma_{2\lambda}$, we shall denote by $(\Gamma_M)_\gamma$ the centralizer of γ in Γ_M. Then $I(f_t,g)$ equals

$$\mathrm{Vol}(\Gamma\cap U\backslash U)\Big\{ \sum_{\gamma\in(\Gamma_{2\lambda}-\{1\})/\Gamma_M} \ \sum_{\delta\in(\Gamma_M)_\gamma\backslash\Gamma_M} f_t(g^{-1}\delta^{-1}\gamma\delta g) \ +$$

$$+ \ \mathrm{Vol}(\Gamma_{2\lambda}\backslash U_{2\lambda})^{-1} \sum_{\gamma\in(\Gamma_\lambda-\{1\})/\Gamma_M} \ \sum_{\delta\in(\Gamma_M)_\gamma\backslash\Gamma_M} \int_{U_{2\lambda}} f_t(g^{-1}\delta^{-1}u\gamma\delta g)\,du \Big\} \ .$$

Insert this expression in (11.41). Let $\mathrm{Re}(s) > 0$. Then, applying Lemma 11.40, we obtain

$$\mathcal{U}(f_t,s) =$$

$$= \mathrm{Vol}(\Gamma\cap U\backslash U)\Big\{ \sum_{\gamma\in(\Gamma_{2\lambda}-\{1\})/\Gamma_M} \mathrm{Vol}((\Gamma_M)_\gamma\backslash M_\gamma) \int_A \int_{M_\gamma\backslash M} f_t(a^{-1}m^{-1}\gamma ma)\cdot$$

$$\cdot\, dm\, a^{-(s\lambda+2\rho)}\,da \ + \tag{11.42}$$

$$+ \ \mathrm{Vol}(\Gamma_{2\lambda}\backslash U_{2\lambda})^{-1} \sum_{\gamma\in(\Gamma_\lambda-\{1\})/\Gamma_M} \mathrm{Vol}((\Gamma_M)_\gamma\backslash M_\gamma) \int_A \int_{M_\gamma\backslash M} \int_{U_{2\lambda}} f_t(ua^{-1}m^{-1}\gamma ma)\cdot$$

$$\cdot\, du\,dm\, a^{-(s+m(\lambda))\lambda}\,da \Big\} \ .$$

Observe that, in view of Lemma 11.40, $\mathrm{Vol}((\Gamma_M)_\gamma\backslash M_\gamma)$ is finite for each $\gamma\in\Gamma_\lambda$ and $\gamma\in\Gamma_{2\lambda}$, respectively. Set $L = MA$ and $\ell = m\oplus a$. To make further progress one has to investigate the orbit structure of the action of L on U_λ (resp. $U_{2\lambda}$) by inner automorphisms. The following result is proved in [51,§1] :

LEMMA 11.43. Let $\alpha\in\{\lambda,2\lambda\}$ and denote by $\mathring{u}_\alpha$ the set of all $X\in u_\alpha$ such that $[X,\ell] = u_\alpha$. Then $\mathring{u}_\alpha$ is a Zariski-open subset of u_α. Moreover, $\mathring{u}_\alpha$ is the disjoint union of finitely many $\mathrm{Ad}_{u_\alpha}(L)$-orbits. The orbits in $u_\alpha - \mathring{u}_\alpha$ are conical. Put $\mathring{U}_\alpha = \exp(\mathring{u}_\alpha)$. Then $\mathring{U}_\alpha$ is the union of finitely many L-conjugacy classes. The M-conjugacy classes, which are contained in $U_\alpha - \mathring{U}_\alpha$ are invariant under conjugation by elements of A.

We observe that u_α is non empty. Indeed we have

$$\Gamma_\alpha - \{1\} \subset \mathring{U}_\alpha \quad , \quad \alpha\in\{\lambda,2\lambda\} . \tag{11.44}$$

Otherwise there would exist $\gamma\in\Gamma_\alpha - \{1\}$ so that the M-conjugacy class $\{\gamma\}_M$ is invariant under conjugation by elements of A. And this implies that

$$\int_A \int_{M_\gamma\backslash M} |f_t(a^{-1}m^{-1}\gamma ma)|\,dm\, a^{-(s\lambda+2\rho)}da =$$

$$= \int_A a^{-(s\lambda+2\rho)}da \int_{M_\gamma\backslash M} |f_t(m^{-1}\gamma m)|\,dm = \infty \quad ,$$

for all $s\in\mathbb{C}$ which contradicts Lemma 11.40.

Another consequence of Lemma 11.43 is that, for each $u\in\mathring{U}_\alpha$ one has $L_u = M_u$. For $u\in U_\alpha$, let

$$C_L(u) = \{l^{-1}ul \mid l\in L\} .$$

Then it follows that, for each $u\in\mathring{U}_\alpha$, one has

$$C_L(u) = A\times M_u\backslash M \quad .$$

It is proved in [51] that M_u is unimodular. Thus the homogeneous space $M_u\backslash M$ carries an invariant measure which we shall denote by $d\dot{m}$. We normalize $d\dot{m}$ as follows. Let $Z\in\mathring{u}_\alpha$ be such that $u = \exp(Z)$.

According to [51], one has $\ell = m_Z \oplus [Z, g_{-\alpha}]$. Therefore, the tangent space of $L_u \backslash L = A \times M_u \backslash M$ at the coset of the identity can be identified with $[Z, g_{-\alpha}]$. Moreover, the restriction of the Killing form to $[Z, g_{-\alpha}]$ is non-degenerate. This induces an invariant measure $d\dot{l}$ on $L_u \backslash L$. Then we assume that $d\dot{m}$ is normalized by $d\dot{l} = dad\dot{m}$. Following Hoffmann [51], we compare the L-invariant measure $dad\dot{m}$ on $C_L(u)$ with the measure du' induced from U. As we have seen above, for each $u' \in C_L(u)$, there exists a unique $a \in A$ so that $u' = a^{-1}m^{-1}uma$ for some $m \in M$. Let us denote this element by $a(u')$. Then the measure $a(u')^{2\rho_\alpha} du'$ is invariant under the action of L by inner automorphisms. Thus there exists a constant $J_\alpha(u) > 0$ such that

$$\int_{C_L(u)} f(u')du' = J_\alpha(u) \int_A \int_{M_u \backslash M} f(a^{-1}m^{-1}uma)d\dot{m}\, a^{-2\rho_\alpha} da \quad , \tag{11.45}$$

for all $f \in C_c(G)$. Observe that J_α satisfies

$$J_\alpha(a^{-1}ua) = a^{-2\rho_\alpha} J_\alpha(u) \quad , \quad a \in A \quad . \tag{11.46}$$

It is easy to describe the function $J_\alpha : \mathring{U}_\alpha \longrightarrow \mathbb{R}^+$ explicitly (c.f. §1 in [51]). Set $j_\alpha = J_\alpha \circ \exp : \mathring{u}_\alpha \to \mathbb{R}^+$. Then $j_\alpha(Z)$ is the Jacobian of the canonical map $L_u \backslash L \longrightarrow C_L(u)$ at the coset of the identity. In particular, the map $Z \longmapsto j_\alpha(Z)^2$ is a polynomial functon on u_α. Set

$$N_\alpha = (J_\alpha)^{1/m(\alpha)} \quad , \quad \alpha \in \{\lambda, 2\lambda\} \quad . \tag{11.47}$$

In view of (11.46), this function satisfies

$$N_\alpha(a^{-1}ua) = a^{-\alpha} N_\alpha(u) \quad , \quad u \in \mathring{U}_\alpha \quad , \quad a \in A \quad . \tag{11.47a}$$

Let $u_1 \in \mathring{U}_\alpha$ and $u \in C_L(u_1)$. Then, using (11.45), we get

$$\begin{aligned} &\int_A \int_{M_u \backslash M} \int_{U_{2\lambda}} f_t(u_2 a^{-1}m^{-1}uma)du_2 d\dot{m}\, a^{-(s\lambda+\rho_\lambda)} da = \\ &= N_\lambda(u)^{-(s+m(\lambda))} \int_{C_L(u_1)} \int_{U_{2\lambda}} f_t(u_2 u_1')du_2\, N_\lambda(u_1')^s du_1' \quad . \end{aligned} \tag{11.48}$$

For $u_2 \in \mathring{U}_{2\lambda}$ and $u \in C_L(u_2)$, we obtain in the same way

$$\int_A \int_{M_u \backslash M} f_t(a^{-1}m^{-1}uma)\,d\dot{m}\, a^{-(s\lambda+\rho)}da =$$

$$= N_{2\lambda}(u)^{-((s+m(\lambda))/2 + m(2\lambda))} \int_{C_L(u_2)} f_t(u_2')N_{2\lambda}(u_2')^{(s+m(\lambda))/2}du_2' \quad . \tag{11.49}$$

Recall that for each $u\in\mathring{U}_\alpha$ one has $N_\alpha(u)>0$. Thus, in view of (11.44), we have $N_\alpha(\gamma) > 0$ for each $\gamma\in\Gamma_\alpha-\{1\}$. Now we introduce the following zeta functions:

For $u_1\in\mathring{U}_\lambda$ and $\mathrm{Re}(s) > 0$ set

$$\zeta_\lambda(s;u_1) = \sum_{\gamma\in(C_L(u_1)\cap\Gamma_\lambda)/\Gamma_M} \frac{\mathrm{Vol}((\Gamma_M)_\gamma\backslash M_\gamma)}{N_\lambda(\gamma)^{s+m(\lambda)}} \quad , \tag{11.50}$$

and for $u_2\in\mathring{U}_{2\lambda}$ and $\mathrm{Re}(s) > 0$ set

$$\zeta_{2\lambda}(s;u_2) = \sum_{\gamma\in(C_L(u_2)\cap\Gamma_{2\lambda})/\Gamma_M} \frac{\mathrm{Vol}((\Gamma_M)_\gamma\backslash M_\gamma)}{N_{2\lambda}(\gamma)^{m(2\lambda)+(s+m(\lambda))/2}} \tag{11.51}$$

In [51] W.Hoffmann proved that the corresponding zeta functions arising from a rank one lattice are uniformly convergent on compacta of $\mathrm{Re}(s)>0$ and admit meromorphic continuations across the line $\mathrm{Re}(s) = 0$. The same method can be used to prove a similar result for the zeta functions (11.50) and (11.51). We shall employ a differend method which is also due to Hoffmann.

<u>LEMMA 11.52</u>. The series $\zeta_\lambda(s;u_1)$ and $\zeta_{2\lambda}(s;u_2)$ defined by (11.50) and (11.51), respectively, are absolutely and uniformly convergent on compacta of $\mathrm{Re}(s) > 0$. $\zeta_\lambda(s;u_1)$ admits a meromorphic continuation to the entire complex plane whose only possible singularities are simple poles at $s=0$ and $s=-2|\rho|/|\lambda|$. Moreover, one has

$$\operatorname*{Res}_{s=0} \zeta_\lambda(s;u_1) = \frac{\mathrm{Vol}(\Gamma_M\backslash M)}{|\lambda|\mathrm{Vol}(\Gamma\cap U)} \quad .$$

$\zeta_{2\lambda}(s;u_2)$ admits a continuation to a holomorphic function in the half-plane $\mathrm{Re}(s) > -m(\lambda)$.

PROOF. Let $\beta \in C_c^\infty(G)$ and set

$$I(\beta,g) = \int_{\Gamma\cap U\backslash U} \sum_{\gamma\in(\Gamma\cap U)-\{1\}} \beta(g^{-1}u^{-1}\,ug)du \quad , \quad g\in G \quad ,$$

$$\mathcal{U}(\beta,s) = \int_A \int_{\Gamma_M\backslash M} I(\beta,ma)dm\, a^{-(s\lambda+2\rho)}da \quad , \qquad s\in\mathbb{C} \quad .$$

Since β has compact support it follows from Lemma 11.37 that there exists a constant $C\in\mathbb{R}$ such that

$$\int_{\Gamma_M\backslash M} I(\beta,ma)dm = 0$$

for $\lambda(\log a) \leq C$. One can now proceed along lines similar to [67,pp.78 -79] , [77,pp.78-79] , and prove that $\mathcal{U}(\beta,s)$ is absolutely and uniformly convergent on compacta of $\mathrm{Re}(s) > 0$ and admits a meromorphic continuation to the entire complex plane, whose only possible singularities are simple poles at $s=0$ and $s=-2|\rho|/|\lambda|$. We leave it to the reader to carry out the details. The proof shows that

$$\mathcal{U}(\beta,s) - \frac{e^{-Cs|\lambda|}}{s|\lambda|}\mathrm{Vol}(\Gamma_M\backslash M)\int_U \beta(u)du -$$

$$- \frac{e^{-C(s|\lambda|+2|\rho|)}}{s|\lambda|+2|\rho|}\mathrm{Vol}(\Gamma_M\backslash M)\,\mathrm{Vol}(\Gamma\cap U\backslash U)\beta(1)$$

is an entire function of s. This implies that the residue of $\mathcal{U}(\beta,s)$ at $s=0$ is given by

$$\frac{\mathrm{Vol}(\Gamma_M\backslash M)}{|\lambda|}\int_U \beta(u)du \quad .$$

We observe that all our computations concerning $\mathcal{U}(f_t,s)$ are valid for $\mathcal{U}(\beta,s)$ as well. One only has to use the fact that $\mathrm{supp}\,\beta$ is compact. We shall now choose β in an appropriate way. Let $\varphi\in C_c^\infty(\mathbb{R}^+)$ and assume that $\varphi > 0$. Let $u_1\in\mathring{U}_\lambda$ and put

$$\psi(u_2'u_1') = \chi_{C_L(u_1)}(u_1')\varphi(J_\lambda(u_1')) \quad .$$

Then it is clear that $\psi\in C^\infty(U)$. Using (2.9), one can extend ψ to a C^∞ function on G. Let $\beta_1\in C_c^\infty(G)$ be such that $\beta_1(u_1) \neq 0$ and put $\beta = \beta_1\psi$. Now apply (11.42), (11.48) and (11.49) with f_t replaced by β and observe that $\beta(u_2) = 0$ for all $u_2\in U_2$. Then, for $\mathrm{Re}(s) > 0$, we obtain

$\mathcal{U}(\beta,s) =$

$$= \mathrm{Vol}(\Gamma\cap U\backslash U)\,\zeta_\lambda(s;u_1)\int_{C_L(u_1)}\Big\{\int_{U_{2\lambda}}\beta_1(u_2u_1')du_2\Big\}\,\varphi(J_\lambda(u_1'))N_\lambda(u_1')^s du_1' \ .$$

This shows that $\zeta_\lambda(s;u_1)$ is absolutely and uniformly convergent on compacta of $\mathrm{Re}(s) > 0$. The integral is obviously an entire function of s and, for a given s, it will be non zero for appropriate choices of β_1 and φ. This gives the meromorphic continuation of $\zeta_\lambda(s;u_1)$ with the claimed position of poles. If we compute the residues at $s=0$ of both sides of the above equation and use the fact that $\beta = \beta_1\varphi$, then we get

$$\frac{\mathrm{Vol}(\Gamma_M\backslash M)}{|\lambda|}\int_U \beta(u)du = \mathrm{Vol}(\Gamma_\lambda\backslash U_\lambda)\ \operatorname*{Res}_{s=0}\zeta_\lambda(s;u_1)\int_{U_\lambda}\int_{U_{2\lambda}}\beta(u_2u_1)du_2du_1 \ .$$

Thus

$$\operatorname*{Res}_{s=0}\zeta_\lambda(s;u_1) = \frac{\mathrm{Vol}(\Gamma_M\backslash M)}{|\lambda|\,\mathrm{Vol}(\Gamma_\lambda\backslash U_\lambda)} \ . \qquad (11.53)$$

Now let $u_2\in\mathring{U}_{2\lambda}$ and put

$$\psi(u_2'u_1') = \chi_{C_L(u_2)}(u_2')\varphi(J_{2\lambda}(u_2')) \ .$$

ψ is a C^∞ function on U. Extend it to a C^∞ function on G. Let $\beta_1\in C_c^\infty(G)$ be such that $\beta_1(u_2)\neq 0$ and set $\beta = \beta_1\psi$. Let $v_1,\dots,v_q\in\mathring{U}_\lambda$ be representatives for the L-conjugacy classes in $\mathring{U}_\lambda$. Using (11.42), (11.48) and (11.49) with f_t replaced by β, we get, for $\mathrm{Re}(s) > 0$,

$$\mathcal{U}(\beta,s) = \mathrm{Vol}(\Gamma\cap U\backslash U)\,\Big\{\mathrm{Vol}(\Gamma_{2\lambda}\backslash U_{2\lambda})^{-1}\sum_{j=1}^{q}\zeta_\lambda(s;v_j)\cdot$$

$$\int_{C_L(v_j)}\int_{U_{2\lambda}}\beta(u_2'u_1')du_2'\ N_\lambda(u_1')^s du_1' \ +$$

$$+\ \zeta_{2\lambda}(s;u_2)\int_{C_L(u_2)}\beta_1(u_2')\varphi(J_{2\lambda}(u_2'))N_{2\lambda}(u_2')^{(s+m(\lambda))/2}du_2'\ \Big\} \ .$$

It is clear that the last integral is an entire function of s and, for a given s, it is non zero for appropriate choices of β_1 and ψ. Furthermore, using the fact that $C_L(v_j) = A\times M_{v_j}\backslash M$ together with (11.47a), it is easy to see that the integrals occurring in the sum are holomorphic for $\mathrm{Re}(s) > -m(\lambda)$. Since we already know that the zeta functions $\zeta_\lambda(s;v_j)$, $j=1,\dots,q$, and $\mathcal{U}(\beta,s)$ are meromorphic functions of s with at most simple poles at $s=0$ and $s=-2|\rho|/|\lambda|$, we get a continuation of $\zeta_{2\lambda}(s;u_2)$ to $\mathrm{Re}(s) > -m(\lambda)$ with at most a simple pole at $s=0$. If we compare the residues at $s=0$ of both sides of the equation above and use (11.53), then it follows that the residue of $\zeta_{2\lambda}(s;u_2)$ at $s=0$ vanishes. Q.E.D.

To complete the first part of the computation of $\mathcal{U}(f_t,0)$ we have to investigate the integrals occurring on the right hand side of (11.48) and (11.49), respectively. It follows from (11.47a) that the function $N_\alpha\circ\exp\colon \mathfrak{u}_\alpha \to \mathbb{R}^+$, $\alpha\in\{\lambda,2\lambda\}$, is homogeneous of degree one. Moreover, as we have seen above, f_t satisfies

$$|f_t(g)| \leqq Ct^{-n/2}\exp\left(-\frac{d^2(gx_o,x_o)}{4t}\right) \quad , \quad g\in G ,$$

$0 < t \leq 1$. This readily implies that both integrals can be analytically continued to the half-plane $\mathrm{Re}(s) > -1$.

Let $C_\lambda(u_1)$ (resp. $C_{2\lambda}(u_2)$) be the constant term of the Laurent expansion at $s=0$ of $\zeta_\lambda(s;u_1)$ (resp. $\zeta_{2\lambda}(s;u_2)$). It is clear that $C_\lambda(u_1)$ (resp. $C_{2\lambda}(u_2)$) depends only on the conjugacy class $C_L(u_1)$ (resp. $C_L(u_2)$). Thus we can consider C_λ (resp. $C_{2\lambda}$) as locally constant functions on $\mathring{U}_\lambda$ (resp. $\mathring{U}_{2\lambda}$). Using (11.42), (11.48), (11.49) and (11.53), we can summarize our results by

$$\mathcal{U}(f_t,0) = \mathrm{Vol}(\Gamma\cap U\backslash U)\Big\{ \int_{U_{2\lambda}} f_t(u)C_{2\lambda}(u)N_{2\lambda}(u)^{m(\lambda)/2}du \quad +$$

$$+ \mathrm{Vol}(\Gamma_{2\lambda}\backslash U_{2\lambda})^{-1}\int_{U_\lambda}\int_{U_{2\lambda}} f_t(u_2u_1)du_2\, C_\lambda(u_1)du_1 \Big\} \quad +$$

$$+ \frac{\mathrm{Vol}(\Gamma_M\backslash M)}{|\lambda|} \int_{U_\lambda}\int_{U_{2\lambda}} f_t(u_2u_1)du_2 \log N_\lambda(u_1)du_1 \qquad .$$

As in [51] , we shall now rewrite this expression in terms of orbital integrals. We begin with some preliminary remarks. Let $Z\in\mathring{\mathfrak{u}}_\lambda$. It follows from [51, Lemma 1] that there exist $Y\in\mathfrak{g}_\lambda$ and $H_\lambda\in\mathfrak{a}$ such that $\{Z,H_\lambda,Y\}$ is a Lie triple. This implies that $\mathrm{ad}(Z)\colon \mathfrak{u}_\lambda \to \mathfrak{u}_{2\lambda}$ is surjective. Indeed, Let $E \subset \mathfrak{g}_{-2\lambda}$ be the orthogonal complement with respect to the Killing form of the image of $\mathrm{ad}(Z)\colon \mathfrak{u}_\lambda \to \mathfrak{u}_{2\lambda}$. Let $W\in E$. Then $\langle \mathrm{ad}(Z)W,V\rangle = 0$ for all $V\in\mathfrak{u}_\lambda$ .Since the Killing form defines a non degenerate pairing between $\mathfrak{g}_{-\lambda}$ and $\mathfrak{g}_\lambda=\mathfrak{u}_\lambda$, it follows that $[Z,W] = 0$. Employing the Jacobi identity, it follows that

$[W,[Z,Y]] = 0$. Hence $0 = [W,H_\lambda] = -2\lambda(H_\lambda)W$ and therefore $E = \{0\}$. Given $Z \in \mathring{\mathfrak{u}}_\lambda$, put $v = \exp(Z)$ and let $\Phi_v : U_v\backslash U \rightarrow U_{2\lambda}$ be defined by $\Phi_v(u) = uvu^{-1}v^{-1}$, $u \in U$. Then this mapping is an isomorphism. We normalize the invariant measures on U_v and $U_v\backslash U$ so that, for $f \in C_c^\infty(U)$, one has

$$\int_{U_v\backslash U}\int_{U_v} f(u\dot{u})du\,d\dot{u} = \int_U f(u)du \quad .$$

There exists a constant $\Delta(v)$ which depends only on the U-conjugacy class of v such that, for $f \in C_c(U_v\backslash U)$, one has

$$\Delta(v) \int_{U_v\backslash U} f(\dot{u})d\dot{u} = \int_{U_{2\lambda}} f(\Phi_v^{-1}(u))du \quad .$$

It follows as in [23,p.85] that

$$\Delta(v) = \frac{\mathrm{Vol}((\Gamma \cap U)_v\backslash U_v)\,\mathrm{Vol}(\Phi_v(\Gamma \cap U)\backslash U_{2\lambda})}{\mathrm{Vol}(\Gamma \cap U\backslash U)} \tag{11.54}$$

Furthermore, we claim that $L_vU_v = P_v$, where $P = LU$. It is easy to see that $L_vU_v \subset P_v$. On the other hand, let $lu \in P_v$, where $l \in L$, $u \in U$. Put $u_1 = uvu^{-1}v^{-1}$. Then we have $lu_1l^{-1}lvl^{-1} = v$. Now recall that the decomposition $U = U_{2\lambda}\cdot U_\lambda$ is invariant under L. Since $v \in U_\lambda$ and $u_1 \in U_{2\lambda}$ it follows that $l \in L_v$ and $u_1 = 1$, i.e. $u \in U_v$ as claimed. Finally, we observe that $P_v = G_v$ (c.f. [51,Lemma 6]). Thus, for each $v \in \mathring{U}_\lambda$, we have

$$G_v = U_vM_v \quad .$$

On G_v we choose the measure $du\,dm$ with the normalizations introduced above. Assume that $v \in \mathring{U}_\lambda$ is such that $N_\lambda(v) = 1$. If we employ the observations above combined with (11.48), then we obtain

$$\int_{C_L(v)}\int_{U_{2\lambda}} f_t(u_2u_1)du_2du_1 =$$

$$= \Delta(v)\int_{L_v\backslash L}\int_{U_v\backslash U} f_t(l^{-1}u^{-1}vul)d\dot{u}\,e^{-2\rho(H(l))}d\dot{l} = \tag{11.55}$$

$$= \Delta(v)\int_{G_v\backslash G} f_t(g^{-1}vg)dg \quad .$$

For the remaining integrals we have to consider $v \in \mathring{U}_{2\lambda}$. In this case we have $P_v = UL_v = UAM_v$. Moreover, one has $G_v = P_v$. Using (11.49), we get

$$\int_{C_L(v)} f_t(u)N_{2\lambda}(u)^{m(\lambda)/2}du = \int_{G_v\backslash G} f_t(g^{-1}vg)dg \quad .$$

In view of (11.47a), in each L-conjugacy class $C_L(u)$, $u\in\mathring{U}_\alpha$, $\alpha\in\{\lambda,2\lambda\}$, one can choose a representative u_1 satisfying $N_\alpha(u_1) = 1$. Let $u_1,\ldots,u_{q_1}\in\mathring{U}_\lambda$ and $v_1,\ldots,v_{q_2}\in\mathring{U}_{2\lambda}$ be representatives for the L-conjugacy classes in $\mathring{U}_\lambda$ and $\mathring{U}_{2\lambda}$, respectively, such that $N_\lambda(u_j) = 1$, $j=1,\ldots,q_1$, and $N_{2\lambda}(v_i) = 1$, $i=1,\ldots,q_2$. Then we get

$$\begin{aligned} \mathcal{U}(f_t,0) = \mathrm{Vol}(\Gamma\cap U\backslash U)\{ &\sum_{i=1}^{q_2} C_{2\lambda}(v_i)\int_{G_{v_i}\backslash G} f_t(g^{-1}v_ig)dg \quad + \\ &+ \mathrm{Vol}(\Gamma_{2\lambda}\backslash U_{2\lambda})^{-1}\sum_{j=1}^{q_1} C_\lambda(u_j)\Delta(u_j)\int_{G_{u_j}\backslash G} f_t(g^{-1}u_jg)dg\ \} \quad + \\ &+ \frac{\mathrm{Vol}(\Gamma_M\backslash M)}{|\lambda|}\int_{U_\lambda}\int_{U_{2\lambda}} f_t(u_2u_1)\log N_\lambda(u_1)\,du_2du_1 \quad . \end{aligned} \tag{11.56}$$

Now recall that the unipotent contribution $\mathcal{U}$ to the L^2-index of D is the constant term in the asymptotic expansion (as $t\to 0$) of $\mathcal{U}(f_t,0)$. Formula (11.56) reduces the investigation of $\mathcal{U}(f_t,0)$ to the study of unipotent orbital integrals and the last integral in (11.56). This is a standard problem in harmonic analysis. One way to study the unipotent orbital integrals is to derive a Fourier inversion formula for the corresponding distributions. Such a Fourier inversion formula does not yet exist in general. For this reason we shall restrict attention to the special case where G is a product $G_1\times\cdots\times G_r$ of groups G_i of real rank one. Under this assumption one can reduce the study of the unipotent orbital integrals occurring in (11.56) to the study of unipotent orbital integrals on the single factors. These orbital integrals can be treated by using the Fourier inversion formula established by D.Barbasch [14]. Observe that the cusps of the manifolds discussed in example 2 at the beginning of Ch.5 are of this type.

For the remainder of this section we shall assume that

$$G = G_1\times\cdots\times G_r \quad ,$$

where each G_i, i=1,...,r, is a connected real semisimple Lie group with finite center and real rank one. Let K_i be a fixed maximal compact subgroup of G_i and let $\mathfrak{k}_i \subset \mathfrak{g}_i$ denote the corresponding Lie algebras, i=1,...,r. Let $u \in G$ be unipotent. Then $u = u_1 \times \cdots \times u_r$, where u_i, i=1,...,r, is a unipotent element of G_i. Moreover, by definition, there exist nilpotent elements $X_i \in \mathfrak{g}_i$ such that $u_i = \exp(X_i)$, i=1,...,r. Let $f_i \in C_c^\infty(G_i)$, i=1,...,r, and set $f = f_1 \otimes \cdots \otimes f_r$. Then $f \in C_c^\infty(G)$ and we have

$$\int_{G_u \backslash G} f(g^{-1}ug)dg = \prod_{i=1}^{r} \int_{(G_i)_{u_i} \backslash G_i} f_i(g_i^{-1}u_i g_i)dg_i \quad . \tag{11.57}$$

The unipotent orbital integrals on G_i can be computed by using the results of [14] . Since X_i is nilpotent there exists a Lie triple $\{X_i, H_i, Y_i\}$ in $\mathfrak{g}_i$ containing X_i, i=1,...,r. The Lie triple $\{X_i, H_i, Y_i\}$ can be normalized so that $X_i - Y_i$ is contained in $\mathfrak{k}_i$. Set $Z_i = X_i - Y_i$. Z_i is a semisimple element of $\mathfrak{g}_i$. Let $n_i = \dim \mathfrak{g}_i$ and $r_i = \dim \mathrm{Cent}_{\mathfrak{g}_i} X_i$. According to [14, Theorem 6.7] , there exists a constant C_{X_i} which depends only on the conjugacy class of X_i such that

$$\lim_{t \to 0^+} t^{(n_i - r_i)/2} \int_{(G_i)_{Z_i} \backslash G_i} f_i(g_i^{-1} \exp(tZ_i) g)dg =$$
$$= C_{X_i} \int_{(G_i)_{u_i} \backslash G_i} f_i(g_i^{-1} u_i g_i)dg_i \quad , \tag{11.58}$$

for each $f_i \in C_c^\infty(G_i)$. The left hand side of (11.58) is closely related to Harish-Chandra's invariant integral. Recall its definition. Let G_o be a connected semisimple Lie group with finite center and let J be a Cartan subgroup of G_o. For each $f \in \mathcal{C}(G_o)$, the invariant integral Φ_f^J relative to J is a function on J' (= regular elements in J) which is defined by

$$\Phi_f^J(j) = \varepsilon_R(j)\Delta(j) \int_{J_o \backslash G_o} f(g^{-1}jg)dg \quad , \quad j \in J' ,$$

(c.f. [78, Vol.II, p.262]). Here J_o denotes the center of J. For the definition of ε_R and Δ see [78, Vol.II, 8.1.1] . Now let $(\mathfrak{g}_i)_{Z_i}$ be the centralizer of Z_i in $\mathfrak{g}_i$ and let $\mathfrak{j}_i$ be a Cartan subalgebra of $(\mathfrak{g}_i)_{Z_i}$ which is fundamental in $(\mathfrak{g}_i)_{Z_i}$. Let J_i be the Cartan subgroup of G_i associated to j_i. If we apply Lemma 9.3.1.8 in [78,Vol.II] to (11.58) and proceed as in the proof of Corollary 6.8 in [14], then it follows that there exists a differential operator Π_{u_i} on J_i such that

$$\lim_{\substack{j\to 1\\ j\in J'}} \Pi_{u_i}\Phi_{f_i}^{J_i}(j) = \int_{(G_i)_{u_i}\backslash G_i} f_i(g^{-1}u_i g)dg \quad . \tag{11.59}$$

Using this result, we get

LEMMA 11.60. Let u be a unipotent element of G. There exist a Cartan subgroup J of G and a differential operator Π_u on J such that, for each $f\in \mathcal{C}(G)$, one has

$$\int_{G_u\backslash G} f(g^{-1}ug)dg = \lim_{\substack{j\to 1\\ j\in J'}} \Pi_u\Phi_f^J(j) \quad .$$

PROOF. It follows from (11.57), (11.59) and a corresponding product formula for the invariant integral that the equation, claimed in the lemma, is true for $f = f_1\otimes\cdots\otimes f_r$, $f_i\in C_c^\infty(G_i)$, $J = J_1\times\cdots\times J_r$, and $\Pi_u = \Pi_{u_1}\times\cdots\times\Pi_{u_r}$. By linearity, this can be extended to all functions $f\in \bigotimes_{i=1}^{r} C_c^\infty(G_i)$. Now observe that, according to Theorem 8.5.1.1 in [78,Vol.II], the mapping $f\longmapsto \Phi_f^J$ of $\mathcal{C}(G)$ into $\mathcal{C}(J')$ is continuous. Moreover, it follows from the results in [71] that the assignment

$$f\longmapsto \int_{G_u\backslash G} f(g^{-1}ug)dg \quad , \quad f\in \mathcal{C}(G) \quad ,$$

is a tempered distribution on G. Thus let $\{f_n\}$ be a sequence of elements in $\bigotimes_{i=1}^{r} C_c^\infty(G)$ such that $f_n\longrightarrow f$ in $\mathcal{C}(G)$, $f\in (G)$. Then we have

$$\sup_{j\in J'} |(\Pi_u\Phi_{f_n}^J)(j) - (\Pi_u\Phi_f^J)(j)| < \varepsilon \quad ,$$

for each $\varepsilon > 0$ and $n \geq n_o(\varepsilon)$. Using this fact we obtain the lemma with $J = J_1\times\cdots\times J_r$ and $\Pi_u = \Pi_{u_1}\times\cdots\times\Pi_{u_r}$. Q.E.D.

COROLLARY 11.61. Assume that $\operatorname{rank} G > \operatorname{rank} K$. Then, for each unipotent element $u\in G$, one has

$$\int_{G_u\backslash G} f_t(g^{-1}ug)dg = 0 \quad .$$

PROOF. According to Lemma 11.60, it is sufficient to prove that for each Cartan subgroup J of G one has $\Phi^J_{f_t} = 0$, unless $\operatorname{rank} G = \operatorname{rank} K$. To prove this fact we shall apply the Fourier inversion formula for the invariant integral established by R.Herb [48] . To describe her result we introduce some notation. For details we refer to [48] . Let J be a Θ-stable Cartan subgroup of G with Lie algebra j . As usually, set $J_K = J \cap K$ and $J_p = \exp(j \cap p)$. Let $C_G(J_p)$ be the centralizer of J_p in G. Then one has $C_G(J_p) = M_1 J_p$, where M_1 is reductive with compact Cartan subgroup J_K. Let $\operatorname{Car}(M_1)$ denote a full set of Θ-stable representatives of M_1-conjugacy classes of Cartan subgroups of M_1. These representatives can be chosen so that, for $B \in \operatorname{Car}(M_1)$, one has $B^o_K \subset J_K$. If $B \in \operatorname{Car}(M_1)$, then $\tilde{B} = BJ_p$ is a Cartan subgroup of G . To each pair (j^*,ν) , with $j^* \in \hat{J}_K$ and $\nu \in j^*_p$, there corresponds a certain tempered invariant distribution $\Theta(J,j^*,\nu)$ on G. If j^* is regular, then $\Theta(J,j^*,\nu)$ is, up to sign, the character of a tempered unitary representation of G induced from a parabolic subgroup of G with split component J_p . Otherwise, $\Theta(J,j^*,\nu)$ is a linear combination of characters of representations which can be embedded in a unitary principal series representation associated to a different class of cuspidal parabolics. Now we can state the main result of [48] given by Theorem 1:

Let $j = j_K j_p$ be a regular element of J. Then

$$\Phi^J_{f_t}(j) = \sum_{B \in \operatorname{Car}(M_1)} \sum_{b^* \in \hat{B}_K} \int_{j^*_p} j_p^{-i\nu'} \int_{b^*_p} C(M_1,B,b^*,\nu,j_K) \cdot$$

$$\cdot \Theta(BJ_p, b^*, \nu' \otimes \nu)(f_t)\,d\nu\,d\nu' \quad ,$$

where the coefficients $C(M_1,B,b^*,\nu,j_K)$ are explicitly computable. Now, if $\operatorname{rank} G > \operatorname{rank} K$, then we have $\dim \mathfrak{a}^* > 0$. Using Proposition 11.26, it follows that $\Phi^J_{f_t} = 0$. Q.E.D.

Thus we can assume that $\operatorname{rank} G = \operatorname{rank} K$. In this case one can apply Theorem 7.1 in [14] to compute the unipotent orbital integrals. First we shall explain some facts we need for this computation. Let $Z_i \in g_i$, $i=1,\dots,r$, be the elements introduced above. Fix a maximal torus T_i in K_i containing the one-parameter group generated by Z_i, and denote by t_i the Lie algebra of T_i. We may assume that the fixed maximal torus T of $K = K_1 \times \cdots \times K_r$ is given by $T = T_1 \times \cdots \times T_r$. Let $L_T \subset \sqrt{-1}\,t^*$ be the lattice which corresponds to the unitary character group $\hat{T}$ of T. Further, let $W(\mathfrak{g}_{\mathbb{C}},\mathfrak{t}_{\mathbb{C}})$ be the Weyl group of $(\mathfrak{g}_{\mathbb{C}},\mathfrak{t}_{\mathbb{C}})$. $\tau\in L_T$ is called regular if $w\tau \neq \tau$ for all $w\neq 1$ in $W(\mathfrak{g}_{\mathbb{C}},\mathfrak{t}_{\mathbb{C}})$. To each $\tau\in L_T$ there is associated a central eigendistribution Θ_τ on G characterized uniquely by certain properties [42,I], [43].

Let Φ_i^+ be the positive roots of the pair $((\mathfrak{g}_i)_{\mathbb{C}},(\mathfrak{t}_i)_{\mathbb{C}})$ chosen as in [87,p.4], and set $\Phi_{Z_i}^+ = \{\beta \in \Phi_i^+ \mid \beta(Z_i) = 0\}$. Put

$$\Phi_u^+ = \Phi_{Z_1}^+ \times \cdots \times \Phi_{Z_r}^+ \quad .$$

Furthermore, let $r_i = \dim \mathrm{Cent}_{\mathfrak{g}_i} X_i$, $s_i = \dim \mathrm{Cent}_{\mathfrak{g}_i} H_i$ and $p_i = r_i - s_i$, $i=1,\dots,r$, where $\{X_i,H_i,Y_i\}$ is the Lie triple containing H_i normalized as above. Finally, let c_{u_i}, $i=1,\dots,r$, be the constant which is given by formula (25) in [14,p.78] and set

$$c_u = c_{u_1} \cdots c_{u_r} \quad .$$

c_u depends only on the conjugacy class of u in G. For each $\tau\in L_T$, put

$$a_u(\tau) = (-1)^l c_u^{-1} \prod_{\beta\in\Phi_u^+} (\tau,\beta) \prod_{i=1}^{r} [\overline{(\tau,Z_i)}]^{p_i} \quad , \tag{11.62}$$

where $l = |\Phi_u^+| + |\Phi_1^+| + \cdots + |\Phi_r^+|$.

<u>PROPOSITION 11.63</u>. Let $u\in G$ be unipotent and assume that $\mathrm{rank}\, G = \mathrm{rank}\, K$. Then one has

$$\int_{G_u\backslash G} f_t(g^{-1}ug)dg = \sum_{\tau\in L_T} a_u(\tau)\,\Theta_\tau(f_t) \quad .$$

<u>PROOF</u>. We use Theorem 7.1 in [14] and proceed in the same way as in the proof of Lemma 11.60 to establish the Fourier inversion formula for the tempered invariant distribution

$$f \longrightarrow \int_{G_u\backslash G} f(g^{-1}ug)dg \quad , \quad f\in \mathcal{C}(G) \; .$$

Then we apply Proposition 11.26. This implies the equation claimed in the proposition. Q.E.D.

Using this proposition we can proceed just as in the proof of Lemma 4.1 in [15]. First, we recall some properties of the distributions Θ_τ, $\tau\in L_T$. Let T' be the set of regular elements in T, W_K the Weyl group of the pair $(\mathfrak{k}_{\mathbb{C}},\mathfrak{t}_{\mathbb{C}})$ and

$$\Delta_T = \prod_{\beta\in\Psi} (e^{\beta/2} - e^{-\beta/2})$$

where Ψ is an appropriately chosen positive root system. Then one has

$$\Theta_\tau|T' = \Delta_T^{-1} \sum_{w\in W_K} \det(w)e^{w\tau} \quad , \tag{11.64a}$$

where e^λ denotes the character of T corresponding to $\lambda\in L_T$. Let $\hat{G}_\tau$ be the set of all $\pi\in\hat{G}$ whose infinitesimal character χ_π is equal to χ_τ. Then

$$\Theta_\tau = \sum_{\pi\in\hat{G}_\tau} \vartheta(\tau,\pi)\Theta_\pi \quad , \tag{11.64b}$$

where $\vartheta(\tau,\pi)\in\mathbb{R}$ and Θ_π is the distributional character of $\pi\in\hat{G}$. Note that, according to [40], $\hat{G}_\tau$ is finite. Moreover, if $\tau\in L_T$ is regular there exists a unique discrete series representation π_τ whose character coincides, up to sign, with Θ_τ. By (11.25) and (11.64b), we get

$$\begin{aligned}\Theta_\tau(f_t) = \sum_\mu n_\mu \sum_{\pi\in\hat{G}_\tau} \vartheta(\tau,\pi)\{\, &\dim(H(\pi)\otimes E_\mu\otimes S^+)^K - \\ &- \dim(H(\pi)\otimes E_\mu\otimes S^-)^K\,\} \; .\end{aligned} \tag{11.65}$$

According to (4.16a) and (4.16c) in [9], $\dim(H(\pi)\otimes E_\mu\otimes S^+)^K - \dim(H(\pi)\otimes E_\mu\otimes S^-)^K$ is equal to the coefficient of $e^{\mu+\rho_c}$ in the Fourier expansion of $(-1)^p\Delta_T\Theta_\pi|T'$, where $p = \frac{1}{2}\dim(G/K)$. Thus in view of (11.64b)

$$\sum_{\pi\in\hat{G}_\tau} \vartheta(\tau,\pi)\{\dim(H(\pi)\otimes E_\mu\otimes S^+)^K - \dim(H(\pi)\otimes E_\mu\otimes S^-)^K$$

coincides with the coefficient of $e^{\mu+\rho_c}$ in the Fourier expansion of $(-1)^p\Delta_T\Theta_\tau|T'$. By (11.64a), this coefficient is 0 if $\mu+\rho_c\notin W_K(-\tau)$ and $(-1)^p\det(w)$ if $\mu+\rho_c = -w\tau$ for some $w\in W_K$. By (11.65) we get

$$\Theta_\tau(f_t) = (-1)^p \sum_{w\in W_K} \det(w)\, n_{-w\tau-\rho_c} \quad .$$

Insert this in the equation of Proposition 11.63. Then we end up with

COROLLARY 11.66. Let $u \in G$ be unipotent. Assume that $\operatorname{rank} G = \operatorname{rank} K$. and let $V = \sum n_\mu E_\mu$ be the unique virtual $k_{\mathbb{C}}$-module determined by (11.24). Then we have

$$\int_{G_u \backslash G} f_t(g^{-1}ug)dg = (-1)^p \sum_\mu n_\mu \sum_{w \in W_K} \det(w) a_u(-w^{-1}(\mu+\rho_c)) .$$

This completes our computation of the unipotent orbital integrals in this particular case.

Assume that $\operatorname{rank} G = \operatorname{rank} K$ and let $u \in G$ be unipotent. For each $\mu \in t^*_{\mathbb{C}}$ set

$$b_u(\mu) = (-1)^p \sum_{w \in W_K} \det(w) a_u(-w^{-1}(\mu+\rho_c)) . \qquad (11.67a)$$

Now we shall introduce certain L-series. As above, let $u_1, \ldots, u_{q_1} \in \mathring{U}_\lambda$ and $v_1, \ldots, v_{q_2} \in \mathring{U}_{2\lambda}$ be representatives for the L-conjugacy classes in $\mathring{U}_\lambda$ and $\mathring{U}_{2\lambda}$, respectively, and assume that $N_\lambda(u_j) = 1$, $j=1,\ldots,q_1$, and $N_{2\lambda}(v_i) = 1$, $i=1,\ldots,q_2$. Then we set

$$L_\lambda(s;\mu) = \sum_{j=1}^{q_1} b_{u_j}(\mu) \Delta(u_j) \zeta_\lambda(s;u_j) \quad , \quad s \in \mathbb{C}$$

$$(11.67b)$$

$$L_{2\lambda}(s;\mu) = \sum_{i=1}^{q_2} b_{v_i}(\mu) \zeta_{2\lambda}(s;v_i) \quad , \quad \operatorname{Re}(s) > -m(\lambda) .$$

Here $\zeta_\lambda(s;u_j)$ and $\zeta_{2\lambda}(s;v_i)$ are the zeta functions defined by (11.50) and (11.51), respectively, and $\Delta(u_j)$ is given by (11.54). By Lemma 11.52, the L-series $L_{2\lambda}(s;\mu)$ is holomorphic in $\operatorname{Re}(s) > -m(\lambda)$. Concerning $L_\lambda(s;\mu)$ we have

LEMMA 11.68. Let $\mu \in t^*_{\mathbb{C}}$ be the highest weight of an irreducible $k_{\mathbb{C}}$-module V_μ and assume that $S^\pm \otimes V_\mu$ lifts to a representation of K. Then $L_\lambda(s;\mu)$ is holomorphic in the half-plane $\operatorname{Re}(s) > -2|\rho|/|\lambda|$

PROOF. It follows from Lemma 11.52 that $L_\lambda(s;\mu)$ is a meromorphic function in the half-plane $\operatorname{Re}(s) > -2|\rho|/|\lambda|$ with at most a simple pole at $s=0$. Since the residue R_λ of $\zeta_\lambda(s;u)$ at $s=0$ is independent of u we get

$$\operatorname*{Res}_{s=0} L_\lambda(s;\mu) = R_\lambda \sum_{j=1}^{q_1} b_{u_j}(\mu) \Delta(u_j) \quad .$$

Since $S_\mu^\pm = S^\pm \otimes V_\mu$ lifts to a representation of K we can define the corresponding homogeneous vector bundles $S_\mu^\pm$ over G/K and the twisted Dirac operator

$$D_\mu^\pm : C^\infty(G/K, S_\mu^\pm) \longrightarrow C^\infty(G/K, S_\mu^\mp)$$

(c.f. [15,p.157]). Let $h_{\mu,t}^\pm : G \longrightarrow \mathrm{End}(S_\mu^\pm)$ be the kernel of the heat semigroup generated by the spinor Laplacian $D_\mu^\mp \circ D_\mu^\pm$ and set $f_{\mu,t} =$ $= \mathrm{tr}\, h_{\mu,t}^+ - \mathrm{tr}\, h_{\mu,t}^-$. Then, by Corollary 11.66, we have

$$b_{u_j}(\mu) = \int_{G_{u_j}\setminus G} f_{\mu,t}(g^{-1}u_j g)dg \quad , \quad j=1,\ldots,q_1 \quad .$$

Using (11.55) and Lemma 11.28, we get

$$\operatorname*{Res}_{s=0} L_\lambda(s;\mu) = R_\lambda \int_U f_{\mu,t}(u)du = 0 \quad .$$

Q.E.D.

In view of Corollary 11.66 the contribution of the unipotent orbital integrals in (11.56) can now be described as follows:

(i) If $\mathrm{rank}\, G > \mathrm{rank}\, K$ this contribution is zero.

(ii) If $\mathrm{rank}\, G = \mathrm{rank}\, K$, let $V = \sum n_\mu E_\mu$ be the unique virtual $k_{\mathbb{C}}$-module determined by (11.24). Then the contribution of the unipotent orbital integrals in (11.56) is given by

$$\sum_\mu n_\mu \{ \mathrm{Vol}(\Gamma_\lambda \setminus U_\lambda) L_\lambda(0;\mu) + \mathrm{Vol}(\Gamma \cap U \setminus U) L_{2\lambda}(0;\mu) \} \quad . \tag{11.69}$$

It remains to investigate the last integral in (11.56). To compute this integral it would be necessary to know the Fourier transform of the tempered distribution

$$f \longrightarrow \int_{U_\lambda} \int_{U_{2\lambda}} f(u_2 u_1)du_2 \log N_\lambda(u_1)du_1 \quad , \quad f \in C(G) \quad ,$$

at least for groups of real rank one. For a group G of real rank one J.Arthur [4] computed the Fourier transform of this distribution on the space of cusp forms $C_o(G)$ of G. This has been completed by G.Warner [88] who computed the Fourier transform on the space of wave packets. Thus the Fourier transform of the distribution above is completely known for real rank one groups. Employing this result we may proceed as in the case of orbital integrals and compute the noninvariant integral occurring in (11.56). However, under a mild additional assumption on the operator D, a simpler method is sufficient to prove that the non-invariant integral in (11.56) actually vanishes. This additional condition is satisfied for geometrically interesting operators.

LEMMA 11.70. Let $G = G_1 \times \cdots \times G_r$ be as above and assume that $r > 1$. Further, assume that the parabolic subgroup P of G which defines the cusp of X is minimal and the operator $D: C^\infty(X,E^+) \longrightarrow C^\infty(X,E^-)$ is such that the matrices $L^\pm$ $\mathrm{End}(V^\pm)$ which are associated to D via (5.4) satisfy $L^\pm = \alpha \mathrm{Id}_{V^\pm}$ for some real number α. Then one has

$$\int_{U_\lambda}\int_{U_{2\lambda}} f_t(u_2u_1)du_2 \log N_\lambda(u_1)du_1 = 0 \quad .$$

PROOF. For each irreducible unitary representation τ of K on V_τ we shall denote by Δ_τ the restriction of $-R(\Omega)\otimes \mathrm{Id}_{V_\tau}$ to the subspace $(L^2(G)\otimes V_\tau)^K$ of $L^2(G)\otimes V_\tau$. Let $p_{\tau,t}: G \longrightarrow \mathrm{End}(V_\tau)$ be the kernel of the heat operator $\exp(-t\Delta_\tau)$. Consider the decomposition

$$V^\pm = \bigoplus_{\tau\in\hat{K}} [\sigma^\pm:\tau]V_\tau$$

of $V^\pm$ into irreducible representations, where $[\sigma^\pm:\tau]$ denotes the multiplicity of the representation τ in $\sigma^\pm$. Using (5.4) and the definition of f_t it follows that

$$f_t = e^{-t\alpha} \sum_{\tau\in\hat{K}} ([\sigma^+:\tau] - [\sigma^-:\tau])\,\mathrm{tr}\, p_{\tau,t} \quad . \tag{11.71}$$

Let χ_τ be the character of τ. Then we have

$$\mathrm{tr}\,\sigma^+ - \mathrm{tr}\,\sigma^- = \sum_{\tau\in\hat{K}} ([\sigma^+:\tau] - [\sigma^-:\tau])\chi_\tau \tag{11.72}$$

in the representation ring $R(K)$. If $\mathrm{rank}\, G > \mathrm{rank}\, K$ then it was established in the proof of (1.2.5) in [15] that V^+ and V^- are isomorphic as K-modules. This implies that $f_t = 0$. Hence we can assume that $\mathrm{rank}\, G = \mathrm{rank}\, K$. Let $s^\pm: k_{\mathbb{C}} \longrightarrow \mathrm{End}(S^\pm)$ be the half-spin representations. If V_μ is an irreducible $k_{\mathbb{C}}$-module with highest weight $\mu \in t^*_{\mathbb{C}}$ such that $S^\pm \otimes V_\mu$ lifts to a representation of K then we let $h^\pm_{\mu,t}: G \longrightarrow \mathrm{End}(S^\pm \otimes V_\mu)$ be the heat kernel for the spinor Laplacian $D^\mp_\mu \circ D^\pm_\mu$ as in the proof of Lemma 11.68. Set $h_{\mu,t} = \mathrm{tr}\, h^+_{\mu,t} - \mathrm{tr}\, h^-_{\mu,t}$.

As above it follows from (1.3.5) in [15] that

$$h_{\mu,t} = \tag{11.73}$$

$$= e^{-t(\|\mu+\rho_c\|^2-\|\rho\|^2)} \sum_{\tau\in\hat{K}} ([s^+\otimes\mu:\tau] - [s^-\otimes\mu:\tau])\operatorname{tr} p_{\tau,t} \ .$$

In view of (11.24) we have

$$\operatorname{tr}\sigma^+ - \operatorname{tr}\sigma^- = \sum_\mu n_\mu(\operatorname{tr}(s^+\otimes\mu) - \operatorname{tr}(s^-\otimes\mu)) =$$

$$= \sum_{\tau\in\hat{K}} \{ \sum_\mu n_\mu([s^+\otimes\mu:\tau] - [s^-\otimes\mu:\tau])\}\chi_\tau \quad .$$

This combined with (11.71) - (11.73) gives

$$e^{t\alpha}f_t = \sum_\mu n_\mu \sum_{\tau\in\hat{K}} ([s^+\otimes\mu:\tau] - [s^-\otimes\mu:\tau])\operatorname{tr} p_{\tau,t} =$$

$$= \sum_\mu n_\mu e^{t(\|\mu+\rho_c\|^2-\|\rho\|^2)} h_{\mu,t} \quad .$$

Thus it is sufficient to prove the lemma for $h_{\mu,t}$. First we shall recall some facts concerning the spin representation. For details we refer to [10] . Let $n=2l$ be even. The group $\mathrm{Spin}(n)$ has a complex representation space S of dimension 2^l called the spin representation. Since $n=2l$ the spin representation is the direct sum of two irreducible subrepresentations $S^\pm$ of dimension 2^{l-1}. If $x_1,\dots,x_l$ are the basic characters of the maximal torus T_o of $SO(2l)$ they can be considered as characters of the maximal torus T of $\mathrm{Spin}(2l)$. Observe that T is a double covering of T_o. The weights of S^+ (resp. S^-) are the characters

$$\tfrac{1}{2}(\pm x_1 \pm x_2 \pm \dots \pm x_l)$$

with an even (resp. odd) number of minus signs. Now let $g_i = k_i \oplus p_i$ be the Cartan decomposition of g_i, $i=1,\dots r$. Since $K = K_1\times\cdots\times K_r$ we have $p = p_1 \oplus \dots \oplus p_r$ and

$$\mathrm{Spin}(p_1)\times\cdots\times\mathrm{Spin}(p_r) \subset \mathrm{Spin}(p) \quad .$$

Let S_i be the $\mathrm{Spin}(p_i)$ representation and S the $\mathrm{Spin}(p)$ representation. Consider S as a representation of $\mathrm{Spin}(p_1)\times\cdots\times\mathrm{Spin}(p_r)$. If one compares the characters of S and $S_1\times\cdots\times S_r$ then it follows that

$$S = S_1 \otimes \cdots \otimes S_r \quad ,$$

considered as representation of $\mathrm{Spin}(p_1) \times \cdots \times \mathrm{Spin}(p_r)$. According to our assumption, we have $\operatorname{rank} G_i = \operatorname{rank} K_i$ for each i, i=1,...,r. This implies that $\dim p_i$ is even and therefore, each spin representation S_i is the direct sum of two irreducible subrepresentations $S_i^{\pm}$. Let $\varepsilon \in \{\pm\}^n$ and set $\operatorname{sign}(\varepsilon) = (-1)^{N(\varepsilon)}$, where $N(\varepsilon) = \#\{i \mid \varepsilon_i = +\}$. If we use the description of $S_i^{\pm}$ and $S^{\pm}$ given above then it follows that

$$S^{\pm} = \bigoplus_{\operatorname{sign}(\varepsilon)=\pm 1} \left[\bigotimes_{i=1}^{r} S_i^{\varepsilon_i} \right] \quad . \tag{11.74}$$

This gives the decomposition of $S^{\pm}$ into irreducible representations of $\mathrm{Spin}(p_1) \times \cdots \times \mathrm{Spin}(p_r)$. Now recall that, via ad, $k_{\mathbb{C}}$ operates on $p_{\mathbb{C}}$. When $p_{\mathbb{C}}$ is endowed with the Killing form, this action becomes skew symmetric

$$\mathrm{ad}\colon k_{\mathbb{C}} \longrightarrow so(p_{\mathbb{C}}) \ .$$

Let

$$s^{\pm}\colon k_{\mathbb{C}} \longrightarrow \mathrm{End}(S^{\pm})$$

be the composition of ad and the half-spin representation $so(p_{\mathbb{C}}) \longrightarrow \mathrm{End}(S^{\pm})$. Since $k = k_1 \oplus \cdots \oplus k_r$ the representation $s^{\pm}$ factors through the representation

$$so(p_{1,\mathbb{C}}) \times \cdots \times so(p_{r,\mathbb{C}}) \longrightarrow \mathrm{End}(S^{\pm})$$

and, in view of (11.74), we get a decomposition of $s^{\pm}$ in subrepresentations. Moreover, the irreducible $k_{\mathbb{C}}$-module V_μ splits in a product

$$V_\mu = V_{\mu_1} \otimes \cdots \otimes V_{\mu_r}$$

of irreducible $k_{i,\mathbb{C}}$-modules V_{μ_i} with highest weight $\mu_i \in t^*_{i,\mathbb{C}}$ and each $S_i^{\pm} \otimes V_{\mu_i}$ lifts to a representation of K_i. Let $S^{\pm}_{\mu_i}$ be the associated vector bundle over G_i/K_i and let

$$D^{\pm}_{\mu_i}\colon C^\infty(G_i/K_i, S^{\pm}_{\mu_i}) \longrightarrow C^\infty(G_i/K_i, S^{\mp}_{\mu_i})$$

be the corresponding twisted Dirac operator (c.f. [15,p.157]).

Furthermore, let $h^{\pm}_{\mu_i,t}: G_i \longrightarrow \mathrm{End}(S_i^{\pm} \otimes V_{\mu_i})$ be the kernel of the heat operator $\exp(-tD^{\mp}_{\mu_i} \circ D^{\pm}_{\mu_i})$. Then it follows from (11.74) that

$$h^{\pm}_{\mu,t} = \sum_{\mathrm{sign}(\varepsilon)=\pm 1} \bigotimes_{i=1}^{r} h^{\varepsilon_i}_{\mu_i,t} \quad .$$

This implies that

$$h_{\mu,t} = \sum_{\varepsilon \in \{\pm\}^n} \mathrm{sign}(\varepsilon) \prod_{i=1}^{r} \mathrm{tr}\, h^{\varepsilon_i}_{\mu_i,t} \quad . \tag{11.75}$$

Since the parabolic subgroup P of G, which occurs in the definition of the cusp of X, is minimal by assumption there exist minimal parabolic subgroups P_i of G_i, $i=1,\dots,r$, such that

$$P = P_1 \times \cdots \times P_r \quad .$$

Let U_i be the unipotent radical of P_i. It is clear that $U = U_1 \times \cdots \times U_r$. Let A_i be a split component of P_i, $i=1,\dots,r$, (c.f. [66,p.32]) and denote the Lie algebra of A_i by $\mathfrak{a}_i$. Since G_i has real rank one we have $\dim \mathfrak{a}_i = 1$. Let λ_i be the unique simple root of $(\mathrm{Lie}(P_i), \mathfrak{a}_i)$. The Lie algebra $\mathfrak{u}_i$ of U_i has the decomposition $\mathfrak{u}_i = \mathfrak{u}_{\lambda_i} \oplus \mathfrak{u}_{2\lambda_i}$. Set $U_{\lambda_i} = \exp(\mathfrak{u}_{\lambda_i})$ and $U_{2\lambda_i} = \exp(\mathfrak{u}_{2\lambda_i})$. $U_{2\lambda_i}$ is the center of U_i and we have $U_i = U_{\lambda_i} \cdot U_{2\lambda_i}$, $U_{\lambda_i} \cap U_{2\lambda_i} = \{1\}$, $i=1,\dots,r$. Since $U_{2\lambda}$ is the center of U it is clear that $U_{2\lambda} = U_{2\lambda_1} \times \cdots \times U_{2\lambda_r}$. Further since $U_\lambda = U/U_{2\lambda}$ and $U_{\lambda_i} = U_i/U_{2\lambda_i}$, $i=1,\dots,r$, we have $U_\lambda = U_{\lambda_1} \times \cdots \times U_{\lambda_r}$. Let $N_{\lambda_i}: U_{\lambda_i} \longrightarrow \mathbb{R}^+$, $i=1,\dots,r$, be the functions defined by (11.47). It is easy to see that

$$N_\lambda(u_1,\dots,u_r) = N_{\lambda_1}(u_1) \cdots N_{\lambda_r}(u_r) \quad , \quad u_i \in U_{\lambda_i} \quad .$$

Using these observations combined with (11.75), we obtain

$$\int_{U_\lambda} \int_{U_{2\lambda}} h_{\mu,t}(uv)\,du \log N_{2\lambda}(v)\,dv =$$

$$= \sum_{\varepsilon \in \{\pm\}^n} \mathrm{sign}(\varepsilon) \sum_{j=1}^{r} \int_{U_{\lambda_j}} \int_{U_{2\lambda_j}} \mathrm{tr}\, h^{\varepsilon_j}_{\mu_j,t}(u_j v_j)\,du_j \log N_{\lambda_j}(v_j)\,dv_j \cdot \tag{11.76}$$

$$\cdot \prod_{i=1}^{r}{}^{(j)} \int_{U_i} \mathrm{tr}\, h^{\varepsilon_i}_{\mu_i,t}(u_i)\,du_i \quad ,$$

where $\Pi^{(j)}$ denotes the product with j-th factor deleted. By Lemma 11.28, we have

$$\int_{U_i} \operatorname{tr} h^+_{\mu_i,t}(u_i)du_i = \int_{U_i} \operatorname{tr} h^-_{\mu_i,t}(u_i)du_i \quad , \quad i=1,\dots,r \ .$$

Since, by our assumption, $r > 1$ this implies that (11.76) vanishes and, as observed above, this is sufficient to prove the lemma. Q.E.D.

REMARK. The Laplace operator acting on differential forms and each twisted Dirac operator satisfy the assumption made in the lemma (compare the discussion after Definition 5.3 and formula (1.3.5) in [15]). Furthermore, assume that $X = \Gamma\backslash G/K$, where Γ is a rank one lattice in G and $D: C^\infty(X,E^+) \longrightarrow C^\infty(X,E^-)$ is a locally invariant elliptic differential operator. Let $V = \sum n_\mu E_\mu$ be the virtual $k_{\mathbb{C}}$-module given by (11.24). Then it follows from (1.2.4) in [15] that

$$L^2\text{-Ind}\, D = \sum_\mu n_\mu L^2\text{-Ind}\, D^+_\mu \quad ,$$

where D^+_μ are the corresponding twisted Dirac operators.

We can now summarize our results by

THEOREM 11.77. Let X be a Riemannian manifold with a cusp of rank one and let $D: C^\infty(X,E^+) \longrightarrow C^\infty(X,E^-)$ be a generalized chiral Dirac operator so that the assumptions of Lemma 11.70 are satisfied. Let the notation be the same as in Theorem 10.32. If $\operatorname{rank} G = \operatorname{rank} K$ let $V = \sum n_\mu E_\mu$ be the unique virtual $k_{\mathbb{C}}$-module given by (11.24). Moreover, for $\mu \in t^*_{\mathbb{C}}$, let $L_\lambda(s;\mu)$ and $L_{2\lambda}(s;\mu)$ be the L-series defined by (11.67b). Then we have

$$L^2\text{-Ind}\, D = \int_X (\alpha^+(z) - \alpha^-(z))dz + \tfrac{1}{2}\eta(0) - \tfrac{1}{2}\operatorname{Tr}(C^+(0)) +$$

$$+ \sum_\mu n_\mu \{ \operatorname{Vol}(\Gamma_\lambda\backslash U_\lambda)L_\lambda(0;\mu) + \operatorname{Vol}(\Gamma \cap U\backslash U)L_{2\lambda}(0;\mu) \} \ ,$$

where it is understood that the last term on the right hand side is zero if $\operatorname{rank} G > \operatorname{rank} K$.

The proof follows immediately by combining Theorem 10.32, (11.34), (11.69) and Lemma 11.70.

REMARK 11.78. It is very likely that similar results will be true without any restriction on G, P and D. First one has to compute the orbital integrals occurring in (11.56). For complex groups this proplem can be treated using the work of D.Barbasch and D.Vogan [80], [81] and R.Hotta and M.Kashiwara [84] . In this case one knows explicit differential operators Π_u, depending on a given unipotent element $u \in G$, such that

$$C_u \int_{G_u \backslash G} f(g^{-1}ug)dg = \lim_{\substack{j \to 1 \\ j \in J'}} [\Pi_u \Phi_f^J](j) \tag{11.79}$$

for a certain constant C_u. Here the limit is taken in a particular chamber depending on u. However, the constant C_u is hard to compute. One can use Theorem 1 in [48] together with Proposition 11.26 to compute the right hand side of (11.78). This implies that the orbital integrals in (11.56) can be expressed as finite linear combinations of discrete series characters Θ_τ evaluated at f_t. However, there are still the noninvariant integrals we have to struggle with in order to get an index formula similar to Theorem 11.77.

CHAPTER XII
THE HIRZEBRUCH CONJECTURE

In order to illustrate our index formula we shall discuss in this chapter one application: The proof of Hirzebruch's conjecture along lines indicated in §6 of [63]. For this purpose let X be the manifold of Example 2 at the beginning of Ch.V. The cusp of this manifold is described in Example 2.3. This description shows that X satisfies the assumption of Theorem 11.77. The dimension of X is $2n$ where n is the degree of the number field F. We introduce an involution τ on the vector bundle Λ^*T^*X by

$$\tau_x \Phi = i^{p(p-1)+n} * \Phi$$

for $\Phi \in \Lambda^p T_x^* X$, $x \in X$. $*$ denotes the Hodge operator [68,V,§1]. For each $x \in X$, let $\Lambda_\pm^* T_x^* X$ be the ± 1 eigenspaces of τ_x. There exist subbundles $\Lambda_\pm^* T^*X$ of Λ^*T^*X with fibre $\Lambda_\pm^* T_x^* X$ at $x \in X$ (c.f. [68,V,§3,Lemma 3]). It is easy to see that the vector bundles Λ^*T^*X are locally homogeneous at infinity. Indeed, let $\mathfrak{g} = \bigoplus_1^n \mathfrak{sl}(2,\mathbb{R})$, $\mathfrak{k} = \bigoplus_1^n \mathfrak{so}(2)$ and $\mathfrak{g} = \mathfrak{k} \oplus \mathfrak{p}$ the corresponding Cartan decomposition. As above we can introduce an involution τ on $\Lambda^* \mathfrak{p}_{\mathbb{C}}^*$. Let $\Lambda_\pm^* \mathfrak{p}_{\mathbb{C}}^*$ be the ± 1 eigenspaces of τ. The representation $\Lambda^* \mathrm{Ad}_{\mathfrak{p}}^* : K \longrightarrow GL(\Lambda^* \mathfrak{p}_{\mathbb{C}}^*)$ decomposes into the direct sum of two subrepresentations

$$\sigma^\pm : K \longrightarrow GL(\Lambda_\pm^* \mathfrak{p}_{\mathbb{C}}^*) \tag{12.1}$$

and the vector bundles $\Lambda_\pm^* T^* H^n$, H the upper half-plane, are associated to $\sigma^\pm$. Let $\Lambda_\pm^*(X)$ be the space of C^∞ sections of the vector bundle $\Lambda_\pm^* T^*X$. Let $\Lambda^*(X)$ be the space of all C^∞ differential forms on X. The involution τ induces a mapping of $\Lambda^*(X)$ into itself which we continue to denote τ. Then $\Lambda_\pm^*(X)$ are the ± 1 eigenspaces of τ. Let $d: \Lambda^*(X) \longrightarrow \Lambda^*(X)$ and $\delta: \Lambda^*(X) \longrightarrow \Lambda^*(X)$ denote the exterior differentiation of forms and its formal adjoint, respectively. One verifies that $d+\delta$ anticommutes with τ [68,V,(17)]. Thus, by restricting $d+\delta$ to $\Lambda_+^*(X)$ we get an operator

$$D_S : \Lambda_+^*(X) \longrightarrow \Lambda_-^*(X) .$$

This is the signature operator. D_S is a first order elliptic differential operator. From its definition it is clear that D_S is locally invariant at infinity. Moreover, the formal adjoint D_S^* of D_S coincides with the restriction of $d+\delta$ to $\Lambda_-^*(X)$. Thus $D_S^*D_S$ (resp. $D_S D_S^*$) is the restriction of the Laplace operator $\Delta = d\delta+\delta d$ to $\Lambda_+^*(X)$ (resp. $\Lambda_-^*(X)$). We shall denote the restriction of Δ to $\Lambda_\pm^*(X)$ by $\Delta_\pm$. By Kuga's lemma [58,p.385], $\Delta_\pm$ coincides with the restriction of the operator $-R(\Omega)\otimes \mathrm{Id}_{\Lambda_\pm^* p_{\mathbb{C}}^*}$ to $(C^\infty(G)\otimes \Lambda_\pm^* p_{\mathbb{C}}^*)^K$ where $G = (SL(2,\mathbb{R}))^n$, $K = (SO(2))^n$. This shows that D_S is a generalized chiral Dirac operator which satisfies the conditions of Theorem 11.77. Thus we can apply Theorem 11.77 to compute the L^2-index of the signature operator D_S. The L^2-norm on $\Lambda^*(X)$ is given by

$$\|\Phi\|^2 = \int_X \Phi \wedge * \bar{\Phi} \quad , \quad \Phi\in\Lambda^*(X) \quad ,$$

and the Hilbert space $L^2\Lambda^*(X)$ is the completion of the space $\Lambda_c^*(X)$ of compactly supported C^∞ differential forms with respect to this norm. Let

$$H_{(2)}^*(X) = \{\Phi\in\Lambda^*(X) \mid \Delta\Phi = 0, \|\Phi\| < \infty\} .$$

This is the space of square integrable harmonic forms on X. Since τ commutes with Δ it maps $H_{(2)}^*(X)$ into itself. Denote by $H_{(2),\pm}^*(X)$ the ± 1 eigenspaces of τ acting on $H_{(2)}^*(X)$. Then it follows from the definition of the L^2-index of D_S that

$$L^2\text{-ind}\, D_S = \dim H_{(2),+}^*(X) - \dim H_{(2),-}^*(X) \quad . \tag{12.2}$$

Now observe that, by definition, τ maps $H_{(2)}^k(X)$ into $H_{(2)}^{2n-k}(X)$, $0 \leq k \leq 2n$. Let $0 \leq k < n$. Then it is clear that

$$H_{(2)}^k(X) \oplus H_{(2)}^{2n-k}(X) \quad \text{and} \quad H_{(2)}^n(X)$$

are invariant under τ. Let $H_{(2),\pm}^k(X)$, $0 \leq k \leq n$, denote the ± 1 eigenspaces of τ acting on these spaces. Then we have

$$H_{(2),\pm}^*(X) = \bigoplus_{0\leq k\leq n} H_{(2),\pm}^k(X) \quad . \tag{12.3}$$

Moreover, if $0 \leq k < n$, then one has

$$H_{(2),\pm}^k(X) = \{\Phi \pm \tau\Phi \mid \Phi \; H_{(2)}^k(X)\} .$$

Thus $\dim \mathcal{H}^k_{(2),+}(X) = \dim \mathcal{H}^k_{(2),-}(X)$, for $k < n$, and, using (12.2) and (12.3), we obtain

$$L^2\text{-Ind}\, D_S = \dim \mathcal{H}^n_{(2),+}(X) - \dim \mathcal{H}^n_{(2),-}(X) \quad . \tag{12.4}$$

There are two cases depending on wether n is even or odd. First assume that $n=2k+1$. In this case the mapping $\tau\colon \mathcal{H}^n_{(2)}(X) \longrightarrow \mathcal{H}^n_{(2)}(X)$ coincides with $i*$. Since $*$ is a real operator one has

$$\tau(\bar{\Phi}) = -\tau(\Phi) \quad , \quad \Phi \in \mathcal{H}^n_{(2)}(X) \quad .$$

This implies that the map $\Phi \longrightarrow \bar{\Phi}$ induces an isomorphism of $\mathcal{H}^n_{(2),+}(X)$ and $\mathcal{H}^n_{(2),-}(X)$ and, by (12.4), we get

$$L^2\text{-Ind}\, D_S = 0 \quad , \quad \text{if} \quad n=2k+1 \ .$$

So we can assume that $n=2k$. Now we have to determine the expression on the right hand side of the index formula of Theorem 11.77. The local signature theorem of [5,§5] implies that

$$\int_X (\alpha^+(z) - \alpha^-(z))dz = \int_X L_k(p_1,\ldots,p_k) \tag{12.5}$$

where $L_k(p_1,\ldots,p_k)$ is the Hirzebruch polynomial in the Pontrjagin forms p_i of the Riemannian manifold X.

Next we shall compute $\eta(0)$. For this purpose we have to determine the restriction $\mathcal{D}_o$ of D_S to the space of U-invariant differential forms on the cusp. It follows from Example 2.3 that $\Gamma_M\backslash X_M$ is a torus of dimension $n-1$ and the cross section N of the cusp is a torus bundle over $\Gamma_M\backslash X_M$ with typical fibre $\Gamma \cap U\backslash U$. The space of harmonic forms on the fibre can be identified with the Lie algebra cohomology $H^*(\mathfrak{u};\mathbb{C})$. Let $\mathbf{H}^*(\mathfrak{u};\mathbb{C})$ be the locally constant sheaf on $\Gamma_M\backslash X_M$ associated to the canonical representation of Γ_M on $H^*(\mathfrak{u};\mathbb{C})$. Now it is easy to see that the space of U-invariant differential forms on the cusp can be identified with the space

$$\Lambda^*(\Gamma_M\backslash X_M \times \mathbb{R}^+, \mathbf{H}^*(\mathfrak{u};\mathbb{C})) \ .$$

of differential forms on $\Gamma_M\backslash X_M \times \mathbb{R}^+$ with values in the local system $\mathbf{H}^*(\mathfrak{u};\mathbb{C})$. The involution τ induces an involution τ_o on this space and we denote the ± 1 eigenspaces of τ_o by

$\Lambda^*_{\pm}(\Gamma_M\backslash X_M \times \mathbb{R}^+, H^*(u;\mathbb{C}))$.

The restriction of this space to $\Gamma_M\backslash X_M$ can be identified with the space $\Lambda^*(\Gamma_M\backslash X_M, H^*(u;\mathbb{C}))$. Let d_M and $*_M$ be the exterior derivative and the Hodge operator on $\Lambda^*(\Gamma_M\backslash X_M, H^*(u;\mathbb{C}))$. Let $C: H^*(u;\mathbb{C}) \longrightarrow H^*(u;\mathbb{C})$ be the operator defined by $C\psi = (2k-n)\psi$ for $\psi \in H^k(u;\mathbb{C})$. Observe that the trace of C is zero. C induces an operator on $\Lambda^*(\Gamma_M\backslash X_M, H^*(u;\mathbb{C}))$ which we denote by C_M. Then a little computation shows that the restriction $\mathcal{D}_o$ of D_S to the space of U-invariant differential forms is given by

$$\mathcal{D}_o = \eta\left(\frac{\partial}{\partial r} + B_M + C_M\right)$$

where

$$B_M\Phi = (-1)^{k+p+1}(\varepsilon *_M d_M - d_M *_M)\Phi$$

and Φ is either a 2p-form ($\varepsilon = 1$) or a (2p-1)-form ($\varepsilon = -1$) contained in $\Lambda^*(\Gamma_M\backslash X_M, H^*(u;\mathbb{C}))$. Thus the operator D_M determined by Lemma 10.3 is given by $D_M = B_M + C_M$. Now recall that $\Delta_M^+ + m^2/4\, I = (D_M + m/2\, I)^2$ where $\Delta_M^+ = -R_{\Gamma_M}(\Omega_M) \otimes \mathrm{Id}$ and Ω_M is given by (4.18). Let $\tilde{D}_M = D_M + m/2\, I$. Since C_M commutes with Δ_M^+ and $\mathrm{Tr}(C) = 0$ we obtain

$$\mathrm{Tr}(\tilde{D}_M e^{-t\tilde{D}_M^2}) = \mathrm{Tr}(B_M e^{-t\Delta_M^+})e^{-(m^2/4)t} + \frac{m}{2}\mathrm{Tr}(e^{-(\Delta_M^+ + m^2/4)}) \quad (12.6)$$

Since $\Gamma_M\backslash X_M$ is a torus it admits an orientation reversing isometry. Therefore the first term on the right hand side vanishes. Furthermore, by the Poisson summation formula one has

$$\mathrm{Tr}(e^{-t\Delta_M^+}) = at^{-(n-1)/2} + O(e^{-c/t})$$

for some $c>0$ and $t \to 0$. Therefore, there is no constant term in the asymptotic expansion of (12.6) as $t \longrightarrow 0$. This implies that the Eta invariant $\eta(0)$ of $\tilde{D}_M$ vanishes.

Let δ_M be the coderivative on $\Lambda^*(\Gamma_M\backslash X_M, H^*(u;\mathbb{C}))$. Then we have $\Delta_M^+ = d_M\delta_M + \delta_M d_M$ and therefore $\Delta_M^+ \geq 0$. Hence the third term in the index formula of Theorem 11.77 vanishes too.

It remains to discuss the L-series. We refer again to Example 2.3 for the description of the cusp. The unipotent radical U of P is commutative and can be identified with $\mathbb{R}^n$. Furthermore, $L = AM$ is isomorphic to $(\mathbb{R}^\times)^n$ and the action of L on U corresponds to the action $y\cdot x = (y_1^2x_1,\dots,y_n^2x_n)$, $y\in(\mathbb{R}^\times)^n$, $x\in\mathbb{R}^n$. $\mathring{U}\subset U$ is defined by $x_1\cdots x_n \neq 0$ and the L-orbits of the action of L on $\mathring{U}$ are parametrized by $\varepsilon\in\{\pm1\}^n$. The orbit $C_L(\varepsilon)\subset\mathring{U}$ which corresponds to $\varepsilon\in\{\pm1\}^n$ is given by $C_L(\varepsilon) = \{x\in\mathbb{R}^n \mid \varepsilon_i x_i > 0,\ i=1,\dots,n\}$. A simple computation shows that the function $J_\lambda\colon \mathbb{R}^n \longrightarrow \mathbb{R}^+$ which is defined by formula (11.45) is given by $J_\lambda(x) = |x_1\cdots x_n|$. Thus, by (11.47), we get $N_\lambda(x) = |x_1\cdots x_n|^{1/n}$. Now recall that there is given a complete $\mathbb{Z}$-module $\mathbf{M}$ in a totally real number field F of degree n and a subgroup $\mathbf{V}\subset U_{\mathbf{M}}^+$ of finite index such that there is an exact sequence

$$0 \longrightarrow \mathbf{M} \longrightarrow \Gamma \longrightarrow \mathbf{V} \longrightarrow 1$$

(c.f. Example 2.3). As in Example 2.3 we identify $\mathbf{M}$ with a lattice in $\mathbb{R}^n$ by sending $\mu \longmapsto (\mu^{(1)},\dots,\mu^{(n)})$. Then we have $\Gamma\cap U \simeq \mathbf{M}$. The group Γ_M is described in Example 2.3. The action of Γ_M on $\Gamma\cap U$ corresponds to the canonical action of $\mathbf{V}$ on $\mathbf{M}$. Nevertheless, Γ need not be equal to the semidirect product of $\mathbf{M}$ and $\mathbf{V}$. Finally, observe that, for each $u\in U$, $M_u = K_M$ which is a finite group. For $\mu\in\mathbf{M}$ set $N(\mu) = \mu^{(1)}\cdots\mu^{(n)}$. For each $\varepsilon\in\{\pm1\}^n$, let $\zeta(s;\varepsilon)$ be the zeta function (11.50) associated to the orbit $C_L(\varepsilon)$. Using the observations above, it follows that

$$\zeta(s;\varepsilon) = \sum_{\substack{\mu\in\mathbf{M}/\mathbf{V}\\ \varepsilon_i\mu^{(i)}>0}} |N(\mu)|^{-s/n+1} \quad . \tag{12.7}$$

Now we can determine the L-series $L_\lambda(s;\tau)$, $\tau\in\mathfrak{t}^*_{\mathbb{C}}$, defined by (11.67b). Since U is commutative we have, for each $u\in U$, $\Delta(u) = 1$ where $\Delta(u)$ is the function (11.54). Next we have to compute the function $b_\varepsilon(\tau)$, $\varepsilon\in\{\pm1\}^n$, which is given by (11.67a). In our case we have $p=\frac{1}{2}\dim H^n = n$ which is even. Moreover, $W_K = \{1\}$. Thus $b_\varepsilon(\tau) = a_\varepsilon(\tau+\rho_c)$. Consider formula (11.62). A representative for the orbit $C_L(\varepsilon)$ is the element $u(\varepsilon)\in U$, whose j-th component is $\begin{pmatrix}1 & \varepsilon_j\\ 0 & 1\end{pmatrix}$. Thus we have $Z_j = \begin{pmatrix}0 & \varepsilon_j\\ -\varepsilon_j & 0\end{pmatrix}$ and this implies $\Phi^+_{u(\varepsilon)} = \emptyset$. Moreover, $l=n$ and $p_i=0$, $i=1,\dots,n$. Thus $a_\varepsilon(\tau+\rho_c) = (c_{u(\varepsilon)})^{-1}$. We have $c_{u(\varepsilon)} = c_{\varepsilon_1}\cdots c_{\varepsilon_n}$, where each c_{ε_j} is given by formula (25) in [14,p.78] . The element Z_o which occurs in this formula coincides with Z_j. The constant d_{Z_j} is determined by Lemma 2.7 in [14] . In the case of $SL(2,\mathbb{R})$ we have $d_{Z_j} = 1$.

The constant C_χ in (25) of [14] is determined by the equation on p.73 in [14]. It follows that $C_\chi = \pi$. Finally, the product over the roots occurring in (25) of [14] is given by $2\sqrt{-1}\,\varepsilon_j$. Thus we obtain $c_{u(\varepsilon)} = \varepsilon_1 \cdots \varepsilon_n (2\pi\sqrt{-1})^n$. Now let us introduce the following L-series associated to $(\mathbf{M},\mathbf{V})$:

$$L(\mathbf{M},\mathbf{V},s) = \sum_{\mu\in(\mathbf{M}-0)/\mathbf{V}} \frac{\operatorname{sign} N(\mu)}{|N(\mu)|^s} \quad , \quad \operatorname{Re}(s) > 1 \ . \tag{12.8}$$

$L(\mathbf{M},\mathbf{V},s)$ has an analytic continuation to a holomorphic function on the whole complex plane (c.f. [63, Lemma 5.56]). Then (12.7) together with the computations above imply that

$$L_\lambda(s;\varepsilon) = \frac{(-1)^k}{(2\pi)^n} L(\mathbf{M},\mathbf{V},\tfrac{s}{n}+1) \quad .$$

Let $S^\pm$ be the half-spin representations of $\mathrm{Spin}(p)$ and set $S = S^+ \oplus S^-$. It is well-known (c.f. [12]) that

$$[\Lambda_+^* p_{\mathbb{C}}^*] - [\Lambda_-^* p_{\mathbb{C}}^*] = [S]([S^+] - [S^-])$$

in the representation ring of $\mathrm{Spin}(p)$ and thus, a fortiori, in $R(k_{\mathbb{C}})$. Thus the virtual $k_{\mathbb{C}}$-module V in Theorem 11.77 is equal to S. Let $S = \sum n_\tau V_\tau$ be the decomposition of S in irreducible $k_{\mathbb{C}}$-modules. Since each V_τ is one-dimensional we have $\sum n_\tau = \dim S = 2^n$. Let

$$d(\mathbf{M}) = \mathrm{Vol}(\mathbb{R}^n/\mathbf{M}) \ .$$

Then our computations show that the last term in the index formula of Theorem 11.77 is equal to

$$\frac{(-1)^k}{\pi^n} d(\mathbf{M}) L(\mathbf{M},\mathbf{V},1) \ . \tag{12.9}$$

This completes our computation of the L^2-index of the signature operator D_S in the case $n=2k$. The non zero contributions are given by (12.5) and (12.9) and the final result is

$$L^2\text{-}\mathrm{Ind}\, D_S = \int_X L_k(p_1,\dots,p_k) + \frac{(-1)^k}{\pi^n} d(\mathbf{M}) L(\mathbf{M},\mathbf{V},1) \ . \tag{12.10}$$

Next we shall identify the L^2-index of D_S with the signature $\mathrm{Sign}(X)$ of X. Since $\Delta_M^+ \geq 0$, it follows from Theorem 4.38 and Theorem 6.17

that the lower bound of the essential spectrum of Δ is positive. This implies that the exterior derivative d on $\Lambda^*(X)$ has closed range as an operator in the Hilbert space of L^2 forms. Therefore, $H^*_{(2)}(X)$ is isomorphic to the L^2-cohomology group $H^*_{(2)}(X;\mathbb{C})$ [89;§1]. Now we may proceed exactly as in [89;§5,§6] to compute the L^2-cohomology. Let W_1 be the interior of X_1 and W_2 the interior of Y_1 where X_1 and Y_1 have the same meaning as in chapter V. Then $X = W_1 \cup W_2$. Since W_1 is smooth and compact, it follows from [83;§4] that there is a long exact sequence

$$\cdots \longrightarrow H^i_{(2)}(X;\mathbb{C}) \longrightarrow H^i_{(2)}(W_1;\mathbb{C}) \oplus H^i_{(2)}(W_2;\mathbb{C}) \longrightarrow H^i_{(2)}(W_1 \cap W_2;\mathbb{C}) \longrightarrow \cdots$$

By [88;(1.6)], we may identify

$$H^i_{(2)}(W_1;\mathbb{C}) \simeq H^i(W_1;\mathbb{C}) \simeq H^i(X;\mathbb{C})$$

$$H^i_{(2)}(W_1 \cap W_2;\mathbb{C}) \simeq H^i(W_1 \cap W_2;\mathbb{C}) \simeq H^i(\partial W_1;\mathbb{C})$$

and the exact sequence may be rewritten as

$$\cdots \longrightarrow H^i_{(2)}(X;\mathbb{C}) \longrightarrow H^i(X;\mathbb{C}) \oplus H^i_{(2)}(W_2;\mathbb{C}) \longrightarrow H^i(\partial W_1;\mathbb{C}) \longrightarrow \cdots$$

Let f_i be the natural mapping

$$f_i\colon H^i_{(2)}(W_2;\mathbb{C}) \longrightarrow H^i(\partial W_1;\mathbb{C}) .$$

The computations on pp.212 - 213 in [89] are also valid in our case and it follows that f_i is an isomorphism whenever $i \leq n-1$, and is the zero mapping when $i \geq n$. Employing the exact sequence above, we obtain that the natural mapping

$$H^i_{(2)}(X;\mathbb{C}) \longrightarrow H^i(X;\mathbb{C})$$

is an isomorphism if $i \leq n-1$, and is injective if $i=n$. Moreover, for $i \geq n$, its image coincides with $H^i_!(X;\mathbb{C})$ - the image of the cohomology with compact supports in $H^i(X;\mathbb{C})$. Hence the natural mapping $H^n_{(2)}(X) \longrightarrow H^n(X;\mathbb{C})$ induces an isomorphism

$$H^n_{(2)}(X) \longrightarrow H^n_!(X;\mathbb{C}) .$$

Finally, observe that, under this isomorphism, the quadratic form

$\omega \longrightarrow (\omega, *\omega)$ on $H^n_{(2)}(X)$ corresponds to the intersection form on $H^n_!(X;\mathbb{C})$ which is given by the cup product. Thus, in view of (12.4), we obtain

$$\mathrm{Sign}(X) = \dim H^n_{(2),+}(X) - \dim H^n_{(2),-}(X) = L^2\text{-}\mathrm{Ind}\, D_S \quad .$$

This result combined with (12.10) gives

PROPOSITION 12.11. Let F be a totally real number field of degree $n=2k$. Let $\mathbf{M}$ be a complete $\mathbb{Z}$-module in F and $\mathbf{V}$ a subgroup of finite index in $U^+_{\mathbf{M}}$. Let X be a Riemannian manifold with a cusp of type $(\mathbf{M},\mathbf{V})$ as defined in Example 2.3. Further, let $L_k(p_1,\ldots,p_k)$ be the Hirzebruch polynomial in the Pontrjagin forms of X, $d(\mathbf{M}) = \mathrm{Vol}(\mathbb{R}^n/\mathbf{M})$ and $L(\mathbf{M},\mathbf{V},s)$ the L-series (12.8). Then one has

$$\mathrm{Sign}(X) = \int_X L_k(p_1,\ldots,p_k) + \frac{(-1)^k}{\pi^n} d(\mathbf{M})L(\mathbf{M},\mathbf{V},1) \quad .$$

Now let us see how this result is related to Hirzebruch's conjecture [50,p.230]. We briefly recall this conjecture. For $d > 0$, let $W(d) = \{z\in H^n \mid \mathrm{Im}(z_1)\cdots\mathrm{Im}(z_n) \geq d\}$. If $d \geq 1$, there is a decomposition $X = X_d \cup Y_d$ where X_d is compact and $Y_d = \Gamma\backslash W(d)$. Let $N = \partial Y_d$ and assume that N is equipped with the orientation induced by the canonical orientation of $\Gamma\backslash W(d)$. Let $z_j = x_j+\sqrt{-1}y_j$, $j=1,\ldots,n$, be the coordinates of H^n. The vector fields $y_j\,\partial/\partial x_j$, $y_j\,\partial/\partial y_j$, $j=1,\ldots,n$, are left-invariant and define a trivialization of TH^n. This trivialization can be pushed down to a trivialization of $T(\Gamma\backslash H^n)$ and this induces a canonical trivialization of $TX_d|N$. Let W be the manifold X_d with orientation reversed. Using the given trivialization of $TW|N$ we get a $SO(2n)$-bundle ξ over W/N. Let $\bar{p}_j\in H^{4j}(W/N;\mathbb{Z})$, $j=1,\ldots,k$, be the Pontrjagin classes of ξ. Then Hirzebruch introduced in [50,§3] the signature defect $\delta(\Gamma)$ associated to Γ by

$$\delta(\Gamma) = L_k(\bar{p}_1,\ldots,\bar{p}_k)[W,N] - \mathrm{Sign}(W) \quad . \tag{12.12}$$

This definition is obviously independent of the choice of $d \geq 1$. The conjecture stated by Hirzebruch in [50,p.230] is the following equality

$$\delta(\Gamma) = \frac{(-1)^k}{\pi^n} d(\mathbf{M})L(\mathbf{M},\mathbf{V},1) \quad .$$

In view of Proposition 12.11, a proof of Hirzebruch's conjecture will follow from

LEMMA 11.13. One has

$$\delta(\Gamma) = \mathrm{Sign}(X) - \int_X L_k(p_1,\dots,p_k) \quad .$$

PROOF. Let $\psi \in C^\infty(\mathbb{R})$ be such that $\psi(r) = 0$ for $r > 3/4$ and $\psi(r)=1$ for $r < 1/4$. For each $d \geq 2$ we introduce $\psi_d \in C^\infty(\mathbb{R})$ by $\psi_d(r) = \psi(r-d+1)$. Let ∇' be the Levi-Cevita connection on X and denote by ∇'' the flat connection on $\Gamma\backslash H^n$ defined by the vector fields $y_j\,\partial/\partial x_j$, $y_j\,\partial/\partial y_j$, $j=1,\dots,n$. For each $d \geq 2$, we introduce on X a new connection by

$$\nabla_d = \psi_d \nabla' + (1 - \psi_d)\nabla'' \quad .$$

Here we consider ψ_d as a function on $N \times [d-1,d]$ and then extend it to a function on X by setting $\psi_d \equiv 0$ on X_{d-1} and $\psi_d \equiv 1$ on Y_d. Let Ω_d be the curvature form of ∇_d and $p_j(\Omega_d)$, $j=1,\dots,k$, the Pontrjagin forms with respect to Ω_d. Since ∇_d is flat on Y_d we have $p_j(\Omega_d) \equiv 0$, $j=1,\dots,k$, in a neighborhood of $N = \partial X_d$. Thus the differential forms $p_j(\Omega_d)$ represent the relative Pontrjagin classes $\bar{p}_j \in H^{4j}(W,N;\mathbb{R})$ of the bundle ξ, defined above, and we have

$$L_k(\bar{p}_1,\dots,\bar{p}_k)[W,N] = - \int_{X_d} L_k(p_1(\Omega_d),\dots,p_k(\Omega_d)) \quad .$$

Recall that W is the manifold X_d with orientation reversed. Since ∇_d coincides on X_{d-1} with the Levi-Cevita connection, we have $p_j(\Omega_d) = p_j$ on X_{d-1}, $j=1,\dots,k$. Now we observe that $L_k(p_1,\dots,p_k)$ vanishes on the cusp Y_1 of X. To see this let d_i be the differential form on H^n which represents the i-th Chern class with respect to the invariant metric on H^n. Let $z_j = x_j + \sqrt{-1}\,y_j$, $j=1,\dots,n$. Then d_i is the i-th elementary symmetric function of the forms

$$\omega_j = - \frac{1}{2\pi} \frac{dx_j \wedge dy_j}{y_j^{\,2}}$$

Since H^n is a complex manifold it follows from Theorem 4.5.1 in [49] that p_j is the j-th elementary symmetric function in the $\omega_j^{\,2}$. But $\omega_j^{\,2} = 0$ and therefore $p_j = 0$, $j > 0$. This shows that $L_k(p_1,\dots,p_k)$ vanishes on $\Gamma\backslash H^n$. This implies that

$$\int_X L_k(p_1,\ldots,p_k) = - L_k(\bar{p}_1,\ldots,\bar{p}_k)[W,N] - \qquad (12.14)$$
$$- \int_{X_d - X_{d-1}} L_k(p_1(\Omega_d),\ldots,p_k(\Omega_d)) \quad .$$

We shall now estimate the second term on the right hand side. For this purpose we compute the curvature form Ω_d with respect to the coordinates $(x_1,y_1,\ldots,x_n,y_n)$ on $Y_1 = \Gamma\backslash W(1)$. Put $w_{2j-1} = x_j$, $w_{2j} = y_j$, $j=1,\ldots,n$. Let ω_j^i be the connection form of ∇_d on Y_1 in the coordinates $(w_1,\ldots,w_{2n})$. A computation shows that

$$\omega_{2j-1}^{2j-1} = -\psi_d \frac{1}{y_j} dy_j - (1-\psi_d)\frac{1}{y_j} dx_j \quad ,$$

$$\omega_{2j}^{2j-1} = -\psi_d \frac{1}{y_j} dx_j \quad , \quad \omega_{2j-1}^{2j} = \psi_d \frac{1}{y_j} dx_j \quad , \quad \omega_{2j}^{2j} = -\frac{1}{y_j} dy_j$$

and $\omega_l^k = 0$ in all other cases.

Recall that $X_M = M^o$ and $M^o = (\mathbb{R}^+)^{n-1}$. Since $\Gamma_M\backslash X_M$ is a flat torus we can introduce coordinates $u_1,\ldots,u_{n-1}$ on $(\mathbb{R}^+)^{n-1}$ so that the set $\{u\in(\mathbb{R}^+)^{n-1} \mid 0 \le u_j < 1,\ j=1,\ldots,n-1\}$ is a fundamental domain for Γ_M and $y_j = r^{1/n}\exp(A_j(u))$, $j=1,\ldots,n$, where $r\in\mathbb{R}^+$ and $A_j(u)$ is a linear function of $u\in(\mathbb{R}^+)^{n-1}$. We shall now use the new coordinates $(r,u_1,\ldots,u_{n-1},x_1,\ldots,x_n)$. If we compute the curvature form Ω_d using the equation $\Omega_j^i = d\omega_j^i - \omega_k^i \wedge \omega_j^k$ then it follows that Ω_j^i is a linear combination of the forms $r^{-1}dr\wedge du_j$, $r^{-(n+1)/n}dr\wedge dx_j$, $r^{-2/n}dx_i\wedge dx_j$, $r^{-1/n}du_i\wedge dx_j$ and $du_i\wedge du_j$ with coefficients which are uniformly bounded as $r\longrightarrow\infty$. Consider a monom

$$\Omega_{i_2}^{i_1}\wedge\Omega_{i_4}^{i_3}\wedge\cdots\wedge\Omega_{i_{2n}}^{i_{2n-1}} \quad .$$

Then this monom can be written as

$$f dr\wedge du_1\wedge\cdots\wedge du_{n-1}\wedge dx_1\wedge\cdots\wedge dx_n$$

with $|f|\le Cr^{-1}$ as $r\to\infty$. Since $L_k(p_1(\Omega_d),\ldots,p_k(\Omega_d))$ is a linear combination of such monoms we get

$$L_k(p_1(\Omega_d),\ldots,p_k(\Omega_d)) = f_d dr\wedge du\wedge dx$$

with $|f_d| \leq Cr^{-1}$ as $r \to \infty$ and C is independent of d. Therefore, the second term on the right hand side of (12.14) tends to zero as $d \to \infty$. But the other two terms are independent of d. Thus we get

$$L_k(\bar{p}_1,\ldots,\bar{p}_k)[W,N] = -\int_X L_k(p_1,\ldots,p_k) \quad .$$

Finally observe that $\mathrm{Sign}(W) = -\mathrm{Sign}(X_d) = -\mathrm{Sign}(X)$. Q.E.D.

COROLLARY 12.15. Let Γ be as above and let $\delta(\Gamma)$ be the signature defect of the cusp singularity of $\Gamma\backslash H^n$. Then one has

$$\delta(\Gamma) = \frac{(-1)^k}{\pi^n} d(\mathbf{M})L(\mathbf{M},\mathbf{V},1) \quad .$$

Thus the proof of the conjecture of Hirzebruch is a consequence of our index theorem. Using Remark 11.78, one can generalize Hirzebruch's conjecture to other locally symmetric spaces of $\mathbb{Q}$-rank one. As soon as the problems discussed in Remark 11.78 are solved, a proof of this generalized Hirzebruch conjecture can be given along the same lines. This will be of interest in connection with the paper [73] .

REFERENCES

[1] ACHIESER, N.I., GLASMAN, I.M.: Theorie der linearen Operatoren im Hilbert Raum. Akademie-Verlag, Berlin 1968.

[2] AMREIN, W.O., PEARSON, D.B., WOLLENBERG, M.: Evanescence of states and asymptotic completeness. Helv. Phys. Acta **53**, 335-351 (1980).

[3] ANDREOTTI, A., VESENTINI, E.: Carleman estimates for the Laplace-Beltrami equation on complex manifolds. Publ. Math. I.H.E.S. **25**, 81-130 (1965).

[4] ARTHUR, J.G.: Some tempered distributions on semi-simple groups of real rank one. Ann. of Math. **100**, 553-584 (1974).

[5] ATIYAH, M.F., BOTT, R., PATODI, V.K.: On the heat equation and the index theorem. Inventiones math. **19**, 279-330 (1973).

[6] ATIYAH, M.F., DONNELLY, H., SINGER, I.M.: Eta invariants, signature defects of cusps, and values of L-functions. Ann. of Math. **118**, 131-177 (1983).

[7] ATIYAH, M.F., PATODI, V.K., SINGER, I.M.: Spectral asymmetry and Riemannian geometry I. Math. Proc. Camb. Phil. Soc. **77**, 43-69 (1975).

[8] ATIYAH, M.F., PATODI, V.K., SINGER, I.M.: Spectral asymmetry and Riemannian geometry III. Math. Proc. Camb. Phil. Soc. **79**, 71-99 (1979).

[9] ATIYAH, M.F., SCHMID, W.: A geometric construction of the discrete series for semisimple Lie groups. Inventiones math. **42**, 1-62 (1977).

[10] ATIYAH, M.F., BOTT, R., SHAPIRO, A.: Clifford modules. Topology **3**, Suppl. 1, 3-38 (1964).

[11] ATIYAH, M.F., SINGER, I.M.: The index of elliptic operators, I. Ann. of Math. **87**, 484-530 (1968).

[12] ATIYAH, M.F., SINGER, I.M.: The index of elliptic operators, III. Ann. of Math. **87**, 546-604 (1968).

[13] BAILY, W., BOREL, A.: Compactifications of arithmetic quotients of bounded symmetric domains. Ann. of Math. **84**, 442-528 (1966).

[14] BARBASCH, D.: Fourier inversion for unipotent invariant integrals Trans. Amer. Math. Soc. **249**, 51-83 (1979).

[15] BARBASCH, D., MOSCOVICI, H.: L^2-index and the Selberg trace formula. J. Functional Analysis **53**, 151-201 (1983).

[16] BAUMGÄRTEL, H., WOLLENBERG, M.: Mathematical scattering theory. Akademie-Verlag, Berlin 1983.

[17] BERGER, M., GAUDUCHON, P., MAZET, E.: Le spectre d'une variété riemannienne. Lecture Notes in Math. **194**. Berlin-Heidelberg-New York, Springer 1971.

[18] BIRMAN, M.S., KREIN, M.G.: On the theory of wave operators and scattering operators. Dokl. Akad. Nauk SSSR **144**, 475-478 (1962).

[19] BOREL, A.: Stable real cohomology of arithmetic groups. Ann. scient. Ec. Norm. Sup., 4e série, **7**, 235-272 (1974).

[20] BOREL, A.: Compact Clifford-Klein forms of symmetric spaces. Topology **2**, 111-122 (1963).

[21] BOREL, A., GARLAND, H.: Laplacian and the discrete spectrum of an arithmetic group. Amer. J. Math. **105**, 309-335 (1983).

[22] COLIN DE VERDIERE, Y.: Une nouvelle démonstration du prolongement méromorphe de séries d'Eisenstein. C. R. Acad. Sci. Paris Ser. I Math., **293**, 361-363 (1981).

[23] DE GEORGE, D.: On a theorem of Osborne and Warner. Multiplicities in the cuspidal spectrum. J. Funct. Analysis **48**, 81-94 (1982).

[24] DODZIUK, J.: Sobolev spaces of differential forms and the De Rham Hodge isomorphism. J. Diff. Geom. **16**, 63-73 (1981).

[25] DONNELLY, H.: Spectral geometry for certain noncompact Riemannian manifolds. Math. Z. **169**, 63-76 (1979).

[26] DONNELLY, H.: Asymptotic expansions for the compact quotients of properly discontinuous group actions. Illinois J. Math. **23**, 485-496 (1979).

[27] DONNELLY, H.: On the cuspidal spectrum for finite volume symmetric spaces. J. Diff. Geom. **17**, 239-253 (1982).

[28] DONNELLY, H.: Eigenvalue estimates for certain noncompact manifolds. Michigan Math. J. **31**, 349-357 (1984).

[29] DUISTERMAAT, J.J., GUILLEMIN, V.W.: The spectrum of positive elliptic operators and periodic bicharacteristics. Invent. Math. **29**, 39-79 (1975).

[30] DUNFORD, N., SCHWARTZ, J.: Linear operators, Part II. New York, London, Interscience 1963.

[31] ENSS, V.: Asymptotic completeness for quantum mechanical potential scattering. Comm. Math. Phys. **61**, 285-291 (1978).

[32] FRIEDMAN, A.: Partial differential equations of parabolic type. Englewood Cliffs, N.J., Printice Hall, 1964.

[33] FRIEDRICHS, K.: Spektraltheorie halbbeschränkter Operatoren und Anwendung auf die Spektralzerlegung von Differentialoperatoren. Math. Ann. **109**, 465-487 (1934).

[34] GILKEY, P.B.: Invariance theory, The heat equation, and the Atiyah-Singer index theorem. Publish or Perish Press, 1985.

[35] GILKEY, P.B.: The residue of the global Eta function at the origin. Advances in Math. **40**, 290-307 (1981).

[36] GREINER, P.: An asymptotic expansion for the heat equation. Arch. Rational Mech. and Anal. **41**, 163-218 (1971).

[37] GROMOV, M., LAWSON, H.B.: Positive scalar curvature and the Dirac operator on complete Riemannian manifolds. Publ. Math. I.H.E.S. **58**, 83-196 (1983).

[38] GUILLOPÉ, L.: Théorie spectrale de particules soumises à courte portée. Séminaire de théorie spectrale et géométrie CHAMBERY-GRENOBLE, 1985-1986, 1-30.

[39] HARDER, G.: On the cohomology of discrete arithmetically defined groups. In: Proc. Intern. Colloq. on Discrete Subgroups of Lie Groups and Application to Moduli, Bombay, 1973, pp. 129-160. Oxford University Press, 1975.

[40] HARISH-CHANDRA: The characters of semisimple Lie groups. Trans. Amer. Math. Soc. **83**, 98-163 (1956).

[41] HARISH-CHANDRA: Invariant eigendistributions on a semisimple Lie algebra. Publ. Math. I.H.E.S. **27**, 5-54 (1965).

[42] HARISH-CHANDRA: Discrete series for semisimple Lie groups. I,Acta Math. **113**, 241-318 (1965), II, Acta Math. **116**, 1-111 (1966).

[43] HARISH-CHANDRA: Two theorems on semisimple Lie groups. Ann. of Math. **83**, 74-128 (1966).

[44] HARISH-CHANDRA: Automorphic forms on semisimple Lie groups. Lecture Notes in Math. **62**, Berlin-Heidelberg-New York, Springer-Verlag, 1968.

[45] HARISH-CHANDRA: Harmonic Analysis on real reductive groups I. The theory of the constant term. J. Funct. Analysis **19**, 104-204 (1975).

[46] HARISH-CHANDRA: Harmonic analysis on real reductive groups III. The Maaß-Selberg relations and the Plancherel formula. Ann. of Math. **104**, 117-201 (1976).

[47] HELGASON, S.: Differential geometry and symmetric spaces. New York-London, Academic Press, 1962.

[48] HERB, R.A.: Discrete series characters and Fourier inversion on semisimple real Lie groups. Trans. Amer. Math. Soc. **277**, 241-262 (1983).

[49] HIRZEBRUCH, F.: Topological methods in algebraic geometry. Third Edition, Springer, Berlin, 1966.

[50] HIRZEBRUCH, F.: Hilbert modular surfaces. Enseignement Math. **19**, 183-281 (1973).

[51] HOFFMANN, W.: The non-semisimple term in the trace formula for rank one lattices. Preprint, Akad. der Wiss. der DDR, Berlin, 1985. (to appear in Crelles Journal)

[52] KATO, T.: Perturbation theory for linear operators. Berlin-Heidelberg-New York, Springer, 1966.

[53] KLINGENBERG, W.: Riemannian geometry. Berlin-New York, de Gruyter 1982.

[54] KOBAYASHI, S., NOMIZU, K.: Foundations of differential geometry. Vol. II. New York, Interscience, 1969.

[55] LANG, S.: $SL_2(\mathbb{R})$. Addison-Wesley, Reading, 1975.

[56] LANGLANDS, R.P.: On the functional equations satisfied by Eisenstein series. Lecture Notes in Math. **544**, Berlin-Heidelberg-New york, Springer , 1976.

[57] LAX, P., PHILLIPS, R.: Scattering theory for automorphic forms. Annals of Math. Studies **87**, Princeton, N.J. 1976.

[58] MATSUSHIMA, Y., MURAKAMI, S.: On vector bundle valued harmonic forms and automorphic forms on symmetric Riemannian manifolds. Ann. of Math. **78**, 365-416 (1963).

[59] MIATELLO, R.J.: Alternating sum formulas for multiplicities in $L^2(\Gamma\backslash G)$, II. Math. Z. **182**, 35-44 (1983).

[60] MILLSON, J.J.: Closed geodesics and the η-invariant. Ann. of Math. **108**, 1-39 (1978).

[61] MOSCOVICI, H.: L^2-index of elliptic operators on locally symmetric spaces of finite volume. In: Operator Algebras and K-Theory, Contemp. Math., Vol. **10**, pp.129-138, 1982.

[62] MÜLLER, W.: Spectral theory for Riemannian manifolds with cusps and a related trace formula. Math. Nachrichten **111**,197-288 (1983).

[63] MÜLLER, W.: Signature defects of cusps of Hilbert modular varieties and values of L-series at s=1. J. Diff. Geom. **20**, 55-119 (1984).

[64] NELSON, E.: Analytic vectors. Ann. of Math. **70**, 572-615 (1959).

[65] OSBORNE, M.S., WARNER, G.: Multiplicities of the integrable discrete series: The case of a nonuniform lattice in an $\mathbb{R}$-rank one semisimple group. J. Funct. Analysis **30**, 287-310 (1978).

[66] OSBORNE, M.S., WARNER, G.: The theory of Eisenstein systems. New York, Academic Press, 1981.

[67] OSBORNE, M.S., WARNER, G.: The Selberg trace formula. I: Γ-rank one lattices. J. reine und angewandte Math. **324**, 1-113 (1981).

[68] PALAIS, R.: Seminar on the Atiyah-Singer index theorem. Annals of Math. Studies **57**, Princeton, N.J. 1965.

[69] PARTHASARATHY, R.: Dirac operators and the discrete series. Ann. of Math. **96**, 1-30 (1972).

[70] POULSEN, N.S.: On C^∞-vectors and intertwining bilinear forms for representations of Lie groups. J. Funct. Analysis **9**, 87-120 (1972).

[71] RAO, R.R.: Orbital integrals in reductive groups. Ann. of Math. **96**, 505-510 (1972).

[72] REED, M., SIMON, B.: Methods of modern mathematical physics. IV: Analysis of Operators. New York, Academic Press, 1978.

[73] SATAKE, I.: On numerical invariants of arithmetic varieties of $\mathbb{Q}$-rank one. In: Automorphic forms of several variables, Progress in Math. Vol. **46**, Boston-Basel-Stuttgart, 1984.

[74] SELBERG, A.: Harmonic analysis and discontinuous groups in weakly symmetric spaces with applications to Dirichlet series. J. Indian Math. Soc. **20**, 47-87 (1956).

[75] TAYLOR, M.: Pseudodifferential operators. Princeton Univ. Press, Princeton, N.J. 1981.

[76] WALLACH, N.R.: An asymptotic formula of Gelfand and Gangolli for the spectrum of $\Gamma\backslash G$. J. Diff. Geom. **11**, 91-101 (1976).

[77] WARNER, G.: Selberg's trace formula for nonuniform lattices: The $\mathbb{R}$-rank one case. Advances in Math. Suppl. Stud. **6**, 1-142 (1979).

[78] WARNER, G.: Harmonic Analysis on Semi-Simple Lie Groups. Vol. I and II, Berlin-New York, Springer, 1972.

[79] WIDDER, D.V.: The Laplace transform. Princeton Univ. Press, Princeton, N.J. 1941.

[80] BARBASCH, D., VOGAN, D.: Primitive ideals and orbital integrals in complex classical groups. Math. Ann. **259**, 153-199 (1982).

[81] BARBASCH, D., VOGAN, D.: Primitive ideals and orbital integrals in complex exceptional groups. J. of Algebra **80**, 350-382 (1983).

[82] BORISOV, N., MÜLLER, W., SCHRADER, R.: Relative index theorems and supersymmetric scattering theory. Preprint 1986.

[83] CHEEGER, J.: On the Hodge theory of Riemannian pseudomanifolds. Proc. Symp. Pure Math. **36**, 91-146. Providence: AMS, 1980.

[84] HOTTA, R., KASHIWARA, M.: The invariant holonomic system on a semisimple Lie algebra. Invent. Math. **75**, 327-358 (1984).

[85] JENSEN, A., KATO, T.: Asymptotic behavior of the scattering phase for exterior domains. Comm. PDE **3**(12), 1165-1195 (1978).

[86] KREIN, M.G.: On the trace formula in the theory of perturbation. Mat. Sbornik **33**(75), 597-626 (1953).

[87] SALLY, P., WARNER, G.: The Fourier transform on semisimple Lie groups of real rank one. Acta Math. **131**, 1-26 (1973).

[88] WARNER, G.: Noninvariant integrals on semisimple groups of $\mathbb{R}$-rank one. J. Funct. Analysis **64**, 19-111 (1985).

[89] ZUCKER, S.: L_2 cohomology of warped products and arithmetic groups. Invent. Math. **70**, 169-218 (1982).

SUBJECT INDEX

(numbers refer to pages)

LIST OF NOTATIONS

(numbers refer to pages)

-***-

Vol. 1090: Differential Geometry of Submanifolds. Proceedings, 1984. Edited by K. Kenmotsu. VI, 132 pages. 1984.

Vol. 1091: Multifunctions and Integrands. Proceedings, 1983. Edited by G. Salinetti. V, 234 pages. 1984.

Vol. 1092: Complete Intersections. Seminar, 1983. Edited by S. Greco and R. Strano. VII, 299 pages. 1984.

Vol. 1093: A. Prestel, Lectures on Formally Real Fields. XI, 125 pages. 1984.

Vol. 1094: Analyse Complexe. Proceedings, 1983. Edité par E. Amar, R. Gay et Nguyen Thanh Van. IX, 184 pages. 1984.

Vol. 1095: Stochastic Analysis and Applications. Proceedings, 1983. Edited by A. Truman and D. Williams. V, 199 pages. 1984.

Vol. 1096: Théorie du Potentiel. Proceedings, 1983. Edité par G. Mokobodzki et D. Pinchon. IX, 601 pages. 1984.

Vol. 1097: R.M. Dudley, H. Kunita, F. Ledrappier, École d'Éte de Probabilités de Saint-Flour XII – 1982. Edité par P.L. Hennequin. X, 396 pages. 1984.

Vol. 1098: Groups – Korea 1983. Proceedings. Edited by A.C. Kim and B.H. Neumann. VII, 183 pages. 1984.

Vol. 1099: C.M. Ringel, Tame Algebras and Integral Quadratic Forms. XIII, 376 pages. 1984.

Vol. 1100: V. Ivrii, Precise Spectral Asymptotics for Elliptic Operators Acting in Fiberings over Manifolds with Boundary. V, 237 pages. 1984.

Vol. 1101: V. Cossart, J. Giraud, U. Orbanz, Resolution of Surface Singularities. Seminar. VII, 132 pages. 1984.

Vol. 1102: A. Verona, Stratified Mappings – Structure and Triangulability. IX, 160 pages. 1984.

Vol. 1103: Models and Sets. Proceedings, Logic Colloquium, 1983, Part I. Edited by G.H. Müller and M.M. Richter. VIII, 484 pages. 1984.

Vol. 1104: Computation and Proof Theory. Proceedings, Logic Colloquium, 1983, Part II. Edited by M.M. Richter, E. Börger, W. Oberschelp, B. Schinzel and W. Thomas. VIII, 475 pages. 1984.

Vol. 1105: Rational Approximation and Interpolation. Proceedings, 1983. Edited by P.R. Graves-Morris, E.B. Saff and R.S. Varga. XII, 528 pages. 1984.

Vol. 1106: C.T. Chong, Techniques of Admissible Recursion Theory. IX, 214 pages. 1984.

Vol. 1107: Nonlinear Analysis and Optimization. Proceedings, 1982. Edited by C. Vinti. V, 224 pages. 1984.

Vol. 1108: Global Analysis – Studies and Applications I. Edited by Yu.G. Borisovich and Yu.E. Gliklikh. V, 301 pages. 1984.

Vol. 1109: Stochastic Aspects of Classical and Quantum Systems. Proceedings, 1983. Edited by S. Albeverio, P. Combe and M. Sirugue-Collin. IX, 227 pages. 1985.

Vol. 1110: R. Jajte, Strong Limit Theorems in Non-Commutative Probability. VI, 152 pages. 1985.

Vol. 1111: Arbeitstagung Bonn 1984. Proceedings. Edited by F. Hirzebruch, J. Schwermer and S. Suter. V, 481 pages. 1985.

Vol. 1112: Products of Conjugacy Classes in Groups. Edited by Z. Arad and M. Herzog. V, 244 pages. 1985.

Vol. 1113: P. Antosik, C. Swartz, Matrix Methods in Analysis. IV, 114 pages. 1985.

Vol. 1114: Zahlentheoretische Analysis. Seminar. Herausgegeben von E. Hlawka. V, 157 Seiten. 1985.

Vol. 1115: J. Moulin Ollagnier, Ergodic Theory and Statistical Mechanics. VI, 147 pages. 1985.

Vol. 1116: S. Stolz, Hochzusammenhängende Mannigfaltigkeiten und ihre Ränder. XXIII, 134 Seiten. 1985.

Vol. 1117: D.J. Aldous, J.A. Ibragimov, J. Jacod, Ecole d'Été de Probabilités de Saint-Flour XIII – 1983. Édité par P.L. Hennequin. IX, 409 pages. 1985.

Vol. 1118: Grossissements de filtrations: exemples et applications. Seminaire, 1982/83. Edité par Th. Jeulin et M. Yor. V, 315 pages. 1985.

Vol. 1119: Recent Mathematical Methods in Dynamic Programming. Proceedings, 1984. Edited by I. Capuzzo Dolcetta, W.H. Fleming and T. Zolezzi. VI, 202 pages. 1985.

Vol. 1120: K. Jarosz, Perturbations of Banach Algebras. V, 118 pages. 1985.

Vol. 1121: Singularities and Constructive Methods for Their Treatment. Proceedings, 1983. Edited by P. Grisvard, W. Wendland and J.R. Whiteman. IX, 346 pages. 1985.

Vol. 1122: Number Theory. Proceedings, 1984. Edited by K. Alladi. VII, 217 pages. 1985.

Vol. 1123: Séminaire de Probabilités XIX 1983/84. Proceedings. Edité par J. Azéma et M. Yor. IV, 504 pages. 1985.

Vol. 1124: Algebraic Geometry, Sitges (Barcelona) 1983. Proceedings. Edited by E. Casas-Alvero, G.E. Welters and S. Xambó-Descamps. XI, 416 pages. 1985.

Vol. 1125: Dynamical Systems and Bifurcations. Proceedings, 1984. Edited by B.L.J. Braaksma, H.W. Broer and F. Takens. V, 129 pages. 1985.

Vol. 1126: Algebraic and Geometric Topology. Proceedings, 1983. Edited by A. Ranicki, N. Levitt and F. Quinn. V, 423 pages. 1985.

Vol. 1127: Numerical Methods in Fluid Dynamics. Seminar. Edited by F. Brezzi, VII, 333 pages. 1985.

Vol. 1128: J. Elschner, Singular Ordinary Differential Operators and Pseudodifferential Equations. 200 pages. 1985.

Vol. 1129: Numerical Analysis, Lancaster 1984. Proceedings. Edited by P.R. Turner. XIV, 179 pages. 1985.

Vol. 1130: Methods in Mathematical Logic. Proceedings, 1983. Edited by C.A. Di Prisco. VII, 407 pages. 1985.

Vol. 1131: K. Sundaresan, S. Swaminathan, Geometry and Nonlinear Analysis in Banach Spaces. III, 116 pages. 1985.

Vol. 1132: Operator Algebras and their Connections with Topology and Ergodic Theory. Proceedings, 1983. Edited by H. Araki, C.C. Moore, Ş. Strătilă and C. Voiculescu. VI, 594 pages. 1985.

Vol. 1133: K.C. Kiwiel, Methods of Descent for Nondifferentiable Optimization. VI, 362 pages. 1985.

Vol. 1134: G.P. Galdi, S. Rionero, Weighted Energy Methods in Fluid Dynamics and Elasticity. VII, 126 pages. 1985.

Vol. 1135: Number Theory, New York 1983–84. Seminar. Edited by D.V. Chudnovsky, G.V. Chudnovsky, H. Cohn and M.B. Nathanson. V, 283 pages. 1985.

Vol. 1136: Quantum Probability and Applications II. Proceedings, 1984. Edited by L. Accardi and W. von Waldenfels. VI, 534 pages. 1985.

Vol. 1137: Xiao G., Surfaces fibrées en courbes de genre deux. IX, 103 pages. 1985.

Vol. 1138: A. Ocneanu, Actions of Discrete Amenable Groups on von Neumann Algebras. V, 115 pages. 1985.

Vol. 1139: Differential Geometric Methods in Mathematical Physics. Proceedings, 1983. Edited by H. D. Doebner and J. D. Hennig. VI, 337 pages. 1985.

Vol. 1140: S. Donkin, Rational Representations of Algebraic Groups. VII, 254 pages. 1985.

Vol. 1141: Recursion Theory Week. Proceedings, 1984. Edited by H.-D. Ebbinghaus, G.H. Müller and G.E. Sacks. IX, 418 pages. 1985.

Vol. 1142: Orders and their Applications. Proceedings, 1984. Edited by I. Reiner and K. W. Roggenkamp. X, 306 pages. 1985.

Vol. 1143: A. Krieg, Modular Forms on Half-Spaces of Quaternions. XIII, 203 pages. 1985.

Vol. 1144: Knot Theory and Manifolds. Proceedings, 1983. Edited by D. Rolfsen. V, 163 pages. 1985.

Vol. 1145: G. Winkler, Choqu
1985.

Vol. 1146: Séminaire d'Algèbr
Proceedings, 1983–1984. Ed
1985.

Vol. 1147: M. Wschebor, Surfaces Aléatoires. VII, 111 pages. 1985.

Vol. 1148: Mark A. Kon, Probability Distributions in Quantum Statistical Mechanics. V, 121 pages. 1985.

Vol. 1149: Universal Algebra and Lattice Theory. Proceedings, 1984. Edited by S. D. Comer. VI, 282 pages. 1985.

Vol. 1150: B. Kawohl, Rearrangements and Convexity of Level Sets in PDE. V, 136 pages. 1985.

Vol 1151: Ordinary and Partial Differential Equations. Proceedings, 1984. Edited by B.D. Sleeman and R.J. Jarvis. XIV, 357 pages. 1985.

Vol. 1152: H. Widom, Asymptotic Expansions for Pseudodifferential Operators on Bounded Domains. V, 150 pages. 1985.

Vol. 1153: Probability in Banach Spaces V. Proceedings, 1984. Edited by A. Beck, R. Dudley, M. Hahn, J. Kuelbs and M. Marcus. VI, 457

Vol. 1165: Seminar on Deformations. Proceedings, 1982/84. Edited by J. Ławrynowicz. IX, 331 pages. 1985.

Vol. 1166: Banach Spaces. Proceedings, 1984. Edited by N. Kalton and E. Saab. VI, 199 pages. 1985.

Vol. 1167: Geometry and Topology. Proceedings, 1983–84. Edited by J. Alexander and J. Harer. VI, 292 pages. 1985.

Vol. 1168: S.S. Agaian, Hadamard Matrices and their Applications. III, 227 pages. 1985.

Vol. 1169: W.A. Light, E.W. Cheney, Approximation Theory in Tensor Product Spaces. VII, 157 pages. 1985.

Vol. 1170: B.S. Thomson, Real Functions. VII, 229 pages. 1985.

Vol. 1171: Polynômes Orthogonaux et Applications. Proceedings, 1984. Edité par C. Brezinski, A. Draux, A.P. Magnus, P. Maroni et A. Ronveaux. XXXVII, 584 pages. 1985.

Vol. 1172: Algebraic Topology, Göttingen 1984. Proceedings. Edited by L. Smith. VI, 209 pages. 1985.

3: H. Delfs, M. Knebusch, Locally Semialgebraic Spaces. XVI,
es. 1985.

4: Categories in Continuum Physics, Buffalo 1982. Seminar.
y F.W. Lawvere and S.H. Schanuel. V, 126 pages. 1986.

Vol. 1175: K. Mathiak, Valuations of Skew Fields and Projective Hjelmslev Spaces. VII, 116pages. 1986.

Vol. 1176: R.R. Bruner, J.P. May, J.E. McClure, M. Steinberger, H_∞ Ring Spectra and their Applications. VII, 388 pages. 1986.

Vol. 1177: Representation Theory I. Finite Dimensional Algebras. Proceedings, 1984. Edited by V. Dlab, P. Gabriel and G. Michler. XV, 340 pages. 1986.

Vol. 1178: Representation Theory II. Groups and Orders. Proceedings, 1984. Edited by V. Dlab, P. Gabriel and G. Michler. XV, 370 pages. 1986.

Vol. 1179: Shi J.-Y. The Kazhdan-Lusztig Cells in Certain Affine Weyl Groups. X, 307 pages. 1986.

Vol. 1180: R. Carmona, H. Kesten, J.B. Walsh, École d'Été de Probabilités de Saint-Flour XIV – 1984. Édité par P.L. Hennequin. X, 438 pages. 1986.

Vol. 1181: Buildings and the Geometry of Diagrams, Como 1984. Seminar. Edited by L. Rosati. VII, 277 pages. 1986.

Vol. 1182: S. Shelah, Around Classification Theory of Models. VII, 279 pages. 1986.

Vol. 1183: Algebra, Algebraic Topology and their Interactions. Proceedings, 1983. Edited by J.-E. Roos. XI, 396 pages. 1986.

Vol. 1184: W. Arendt, A. Grabosch, G. Greiner, U. Groh, H.P. Lotz, U. Moustakas, R. Nagel, F. Neubrander, U. Schlotterbeck, One-parameter Semigroups of Positive Operators. Edited by R. Nagel. X, 460 pages. 1986.

Vol. 1185: Group Theory, Beijing 1984. Proceedings. Edited by Tuan H.F. V, 403 pages. 1986.

Vol. 1186: Lyapunov Exponents. Proceedings, 1984. Edited by L. Arnold and V. Wihstutz. VI, 374 pages. 1986.

Vol. 1187: Y. Diers, Categories of Boolean Sheaves of Simple Algebras. VI, 168 pages. 1986.

Vol. 1188: Fonctions de Plusieurs Variables Complexes V. Séminaire, 1979–85. Edité par François Norguet. VI, 306 pages. 1986.

Vol. 1189: J. Lukeš, J. Malý, L. Zajíček, Fine Topology Methods in Real Analysis and Potential Theory. X, 472 pages. 1986.

Vol. 1190: Optimization and Related Fields. Proceedings, 1984. Edited by R. Conti, E. De Giorgi and F. Giannessi. VIII, 419 pages. 1986.

Vol. 1191: A.R. Its, V.Yu. Novokshenov, The Isomonodromic Deformation Method in the Theory of Painlevé Equations. IV, 313 pages. 1986.

Vol. 1192: Equadiff 6. Proceedings, 1985. Edited by J. Vosmansky and M. Zlámal. XXIII, 404 pages. 1986.

Vol. 1193: Geometrical and Statistical Aspects of Probability in Banach Spaces. Proceedings, 1985. Edited by X. Femique, B. Heinkel, M.B. Marcus and P.A. Meyer. IV, 128 pages. 1986.

Vol. 1194: Complex Analysis and Algebraic Geometry. Proceedings, 1985. Edited by H. Grauert. VI, 235 pages. 1986.

Vol.1195: J.M. Barbosa, A.G. Colares, Minimal Surfaces in $\mathbb{R}^3$. X, 124 pages. 1986.

Vol. 1196: E. Casas-Alvero, S. Xambó-Descamps, The Enumerative Theory of Conics after Halphen. IX, 130 pages. 1986.

Vol. 1197: Ring Theory. Proceedings, 1985. Edited by F.M.J. van Oystaeyen. V, 231 pages. 1986.

Vol. 1198: Séminaire d'Analyse, P. Lelong – P. Dolbeault – H. Skoda. Seminar 1983/84. X, 260 pages. 1986.

Vol. 1199: Analytic Theory of Continued Fractions II. Proceedings, 1985. Edited by W.J. Thron. VI, 299 pages. 1986.

Vol. 1200: V.D. Milman, G. Schechtman, Asymptotic Theory of Finite Dimensional Normed Spaces. With an Appendix by M. Gromov. VIII, 156 pages. 1986.